Climate and Energy:

The Feasibility of Controlling
CO_2 Emissions

The Dutch Ministry of Housing, Physical Planning and Environment has asked

the National Institute of Public Health and Environmental Protection (RIVM)

and

the Netherlands Energy Research Foundation (ECN)

to review the relevant expertise in The Netherlands with respect to options for CO_2 abatement at the national and a global scale.

The results of this study can be found in this book, the Editors of which can be contacted at the following addresses:

P.A. Okken
Netherlands Energy
Research Foundation
(ECN)
P.O. Box 1
1755 ZG Petten (NH)
The Netherlands

R.J. Swart
National Institute of
Public Health and
Environmental
Protection (RIVM)
P.O. Box 1
3720 BA Bilthoven
The Netherlands

S. Zwerver
National Institute of
Public Health and
Environmental
Protection (RIVM)
P.O. Box 1
3720 BA Bilthoven
The Netherlands

Full addresses of Authors can be found in their respective Chapters.

Climate and Energy:

The Feasibility of Controlling CO$_2$ Emissions

edited by

P. A. Okken
Netherlands Energy Research Foundation,
Petten (NH), The Netherlands

R. J. Swart
National Institute of Public Health
and Environmental Protection,
Bilthoven, The Netherlands

and

S. Zwerver
National Institute of Public Health
and Environmental Protection,
Bilthoven, The Netherlands

KLUWER ACADEMIC PUBLISHERS
DORDRECHT / BOSTON / LONDON

ISBN-13: 978-94-010-6704-1   e-ISBN-13: 978-94-009-0485-9
DOI: 10.1007/978-94-009-0485-9

Published by Kluwer Academic Publishers,
P.O. Box 17, 3300 AA Dordrecht, The Netherlands.

Kluwer Academic Publishers incorporates
the publishing programmes of
D. Reidel, Martinus Nijhoff, Dr W. Junk and MTP Press.

Sold and distributed in the U.S.A. and Canada
by Kluwer Academic Publishers,
101 Philip Drive, Norwell, MA 02061, U.S.A.

In all other countries, sold and distributed
by Kluwer Academic Publishers Group,
P.O. Box 322, 3300 AH Dordrecht, The Netherlands.

printed on acid free paper

CONTENTS

1. PREFACE   1

    P. Vellinga

2. SUMMARY AND INTRODUCTION   3

    P.A. Okken, R.J. Swart and S. Zwerver

3. IMAGE: A TOOL FOR LONG TERM GLOBAL GREENHOUSE   18
    POLICY ANALYSIS

    R.J. Swart and J. Rotmans

4. HOW TO DECREASE THE CO_2 EMISSIONS WHILE SAVING MONEY   28

    J. Bosma

5. ENERGY CONSERVATION FOR A LONG TERM, SUSTAINABLE   41
    ENERGY POLICY

    H.Y. Becht and J.P. van Soest

6. PROSPECTS FOR CARBON DIOXIDE EMISSION REDUCTION   60

    F.M.J.A. Diepstraten and W. van Gool

7. THE POTENTIAL OF RENEWABLE ENERGY TO REDUCE CO_2 78
    EMISSIONS

    E.H. Lysen

8. THE PROSPECTS OF PHOTOVOLTAIC SOLAR ENERGY   95
    CONVERSION

    W.C. Turkenburg, E.A. Alsema and K. Blok

9. REFORESTATION, A FEASIBLE CONTRIBUTION TO   107
   REDUCING THE ATMOSPHERIC CARBON DIOXIDE CONTENT?

   K.F. Wiersum and P. Ketner

10. THE RECOVERY OF CARBON DIOXIDE FROM POWER PLANTS   125

    C.A. Hendriks, K. Blok and W.C. Turkenburg

11. STORAGE OF CARBON DIOXIDE IN THE OCEANS   143

    H.J.W. de Baar and M.H.C. Stoll

12. DISPOSAL OF CARBON DIOXIDE IN DEPLETED NATURAL GAS   178
    RESERVOIRS

    A.C. van der Harst and A.J.F.M. van Nieuwland

13. AN ALL-ELECTRIC SOCIETY FOR LESS CO_2?   189

    J.H.C. van der Veer and T.W.M. van Wunnik

14. HYDROGEN AS AN ENERGY CARRIER TO REDUCE CO_2 200
    EMISSIONS

    H. Muis and A.F.L. Slob

15. THE CARBON-DIOXIDE SUBSTITUTION POTENTIAL OF   216
    METHANE AND URANIUM RESERVES

    H.J.M. de Vries

16. INTEGRATED ASSESSMENT OF ENERGY-OPTIONS FOR CO_2 243
    REDUCTION

    T. Kram and P.A. Okken

17. CARBON DIOXIDE AND POLICY OPTIONS   261

    W. van Gool

Contributions:

E.A. Alsema

H.J.W. de Baar

H.Y. Becht

K. Blok

J. Bosma

F.M.J.A. Diepstraten

W. van Gool

A.C. van der Harst

C.A. Hendriks

P. Ketner

T. Kram

E.H. Lysen

H. Muis

A.J.F.M. van Nieuwland

P.A. Okken

J. Rotmans

A.F.L. Slob

J.P. van Soest

M.H.C. Stoll

R.J. Swart

W.C. Turkenburg

J.H.C. van der Veer

P. Vellinga

H.J.M. de Vries

K.F. Wiersum

A.W.M. van Wunnik

S. Zwerver

Translations:

mw. A.J. Blommesteijn
mw. W. Browne

Text processing:

mw. B. van den Berg
mw. J. Britstra-Sunnotel

PREFACE

Rapidly increasing concentrations of greenhouse gases in the atmosphere, emerging evidence of global warming and the threat of uncontrollable climate feedback mechanisms are now triggering international action to reduce the emissions of greenhouse gases.

In 1989 the Intergovernmental Panel on Climate Change (IPCC), established by the United Nations Environment Programme and the World Meteorological Organization, started preparations for an international convention on climate. This convention is to be followed by protocols (agreements) on the reduction of the emissions of greenhouse gases and other measures and implementation mechanisms to preserve the global climate.

After the CFC's, CO_2 is the next in line, as the sources and abatement measures for CH_4 and N_2O are as yet insufficiently understood. However, the abatement of CO_2 .is a far reaching issue. It will require major changes within the most important sectors of the economy: energy (production and use) and agriculture (deforestations and land use patterns). Given this situation it is not so surprising that national governments are hesitant to take action. One reason is the remaining uncertainty regarding the rate and the extent of climate change. However, further analysis will show that the uncertainties will be outweighed by the increasing risks when measures to reduce the emission of greenhouse gases are delayed. A second reason for a hesitant approach may well be the lack of knowledge on the possibilities of CO_2 abatement measures and on the unpredictability of the consequences of such measures.

This situation calls for a prudent, step by step appraoch whilst maintaining long term flexibility. As a first step, measures should be taken that are beneficial not only for delaying climate change but for other environmental and economic purposes as well.

Along with a number of other nations, The Netherlands has taken such a first step. In the National Environmental Policy Plan of 1989 a freeze of CO_2 emissions has been announced at a level of 1989/1990, to be reached before the year 2000. In view of the reduction of emissions ultimately required to stabilize atmospheric concentrations this is a very limited step. However, in view of the present rate of growth of CO_2 emissions it will take a considerable effort. Meanwhile further reduction steps can be investigated.

To generate ideas for implementing the first step and to investigate the possibilities for further steps, the Ministry of Housing, Physical Planning and Environment has asked the National Institute of Public Health and Environmental Protection and the Netherlands Energy Research Foundation to review the relevant expertise in The Netherlands with respect to options for CO_2 abatement at the national and a global scale.

In preparation for the Ministerial Conference on Atmospheric Pollution and Climate Change to be held on November 6 and 7 in The Netherlands, the reviewed and developed options have been written up and included in this report. The options have been presented by the authors at a symposium specifically devoted to the subject, that was held in Utrecht on September 27, 1989.

This report is written for all those involved in the decision making process regarding climate change policy (i.e. scientists, technicians, press, non-governmental organizations, policy makers in the field of industry, agriculture, forestry and environment, and politicians).

The report presents a series of promising options for CO_2 abatement from a technological point of view. In view of the wide range of possibilities, the reader may well conclude that the social and institutional hurdles to be taken to implement existing technology require as much effort as the generation of new technologies.

On behalf of the authors I invite the reader to comment on the various options presented in this report. I also challenge other nations to present their expert views in a similar way, mobilizing a maximum of creative thinking and developing a broad basis for a well analyzed and a well planned approach to the preservation of the global climate.

Pier Vellinga
Coordinator National Climate Change Programme
Vice chairman of IPCC Response Strategies Working Group

Ministry of Housing, Physical Planning and Environment,
The Netherlands

SUMMARY AND INTRODUCTION

P.A. Okken
Netherlands Energy Research Foundation
R.J. Swart and S. Zwerver
National Institute of Public Health and
Environmental Protection

1. THE CLIMATE SYSTEM

The global environment has become a very important issue in the last few years. In te past decade all over the world the belief and fear have increased that life on earth, as we know it, is threatened by the activities of mankind. The greenhouse effect is essential for life on earth. For million of years this effect has made life on earth possible. Carbon dioxide (CO_2) and water vapour in the atmosphere allow the short wave radiation of the sun's spectrum to pass through to the earth. The re-radiated long light waves in the infrared domain, however, are absorbed on their way back from the earth. This way the absorbing gases hold back the heat which would otherwise be lost into space. The warming of the atmosphere produces an average temperature of the earth's surface of +15°C instead of the -18°C it would have if the earth's atmosphere would only contain nitrogen (N_2) and oxygen (O_2). Through this process of transmission and absorption greenhouse gases have an important function in the regulation of the temperature on earth.
  During the past 200 years the use of fossil fuels has made mankind a dominating factor in the greenhouse effect. The large-scale and worldwide use of fossil fuels to generate heat, power, and electricity for a growing population as well as the burning of large areas of tropical forest for agriculture have given an enormous emission of CO_2 into the atmosphere. For more than a century, ever since the beginning of the industrial revolution, the use of fossil fuels has been increasing and this development still continues. Measurements have shown the CO_2 concentration in the atmosphere to increase. This concentration is now 25% higher than in 1800. If the current trend continues, the CO_2 concentration will have doubled towards the middle of

3

the next century. Apart from CO_2 other greenhouse gases are also emitted as a result of human activities. The most important are methane (CH_4), chlorofluorocarbons, especially CFC 11 and CFC 12, nitrous oxide (N_2O) and ozone (O_3). The concentrations of some of these trace gases (for example methane and chlorofluorocarbons) have increased by even larger factors.

On the basis of the results of global circulation models there is a growing consensus among scientists all over the world that the doubling of CO_2 in the atmosphere might lead to an increase in the average temperature of 1.5° to 4.5°, an increase in precipitation of 7 to 15% and a rise of the sea level of 10 to 100 cm. These are only averages. The increase in temperature might be greater at the poles and less at the equator for instance. Around the middle of the next century, which is in approximately 60 years, this double CO_2 concentration is expected to be reached. The process of climatic change will even be enchanced and accelerated by the influence of the other greenhouse gases. If the expected rise in temperature for the next 60 years is compared to the global rise in temperature of roughly 5°C after the last glacial period 15,000 to 50,000 years ago, one realizes how serious the climatic change will be and how dramatically this change could effect the environment of the world. One must assume that not only the impact on climate and environment, but also the impact on our society will be very great indeed.

Climate change will have a direct impact in many different fields such as agriculture crops, forestry, floods, salt water penetration, water supply, natural ecosystems, and the spread of infectuous diseases. But the economy and society will also be indirectly affected owing to the effects on the food supply, the world timber production and energy supply. Especially for the rapidly increasing population of the Third World, where ecosystems are often vulnerable, the risks are very high. However, also in a highly developed country with a complex social and economic structure, the effects on the environment and the economy may be profound, as has been demonstrated most clearly perhaps by the drought in the Mid-West of the U.S.. Exhaustion of agricultural land and water supplies may lead to political instability and may cause a refugee problem. The geopolitical impact of all this can only be guessed.

2. THE ENERGY SECTOR

Apart from CO_2 and nitrogen compounds (N_2O, NO_x), the exploration and use of fossil fuels causes emissions of carbon monoxide (CO) and hydrocarbons (such as methane). The energy sector's contribution to the potential climate change caused by the greenhouse effect is 75%, the remaining 25% being caused by agriculture, industry and miscellaneous human activities. The share of CO_2 in this climatic change is 50%. The key to solving this worldwide environmental problem has therefore to be found in the field of energy supply by fossil fuels and to a lesser extent in the field of deforestation.

In the past energy use and economic growth were closely connected. After the energy crises in the seventies this connection became less close. Low energy prices and increasing economic growth, however, have recently caused energy consumption to increase again. The energy system of the world is based on carbonaceous fuels (about 80%) and biomass (about 15%). The lack of data makes it difficult to quantify the contribution of non-commercial sources of energy (biomass). As a result of regrowth the use of biomass for energy purpose will only partly contribute to the net CO_2 emissions. Also the percentage of non-commercial energy use is likely to decline. The emissions of CO_2 from fossil fuels are distributed unevenly over the world (see figure 1).

mln ton $C\bar{O}2$

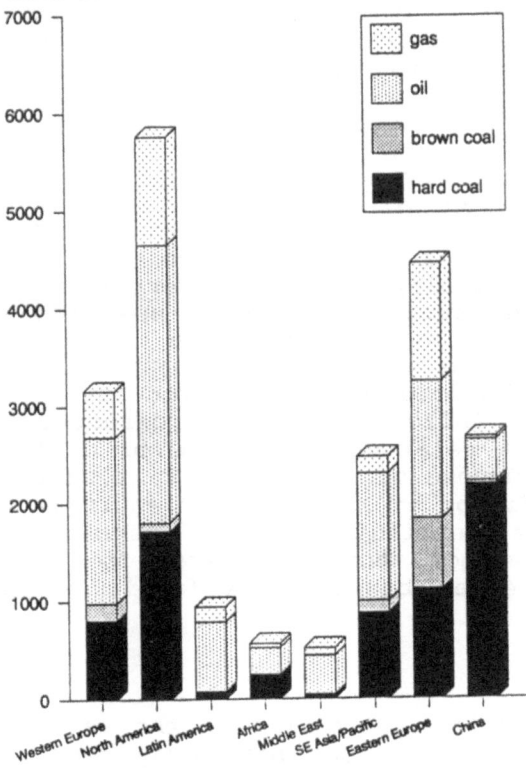

Figure 1

Quantification of present and future emissions of trace gases from fossil fuel combustion is a basic step in the designing of climate change response strategies. The emissions of carbon dioxide are reasonably well-known, since they can be directly related to the consumption of the major fossil fuels, on which there are many data. Uncertainties about precise CO_2 emission factors are small. They arise from uncertainties about consumption, for instance by the military sector, from the variability of certain emission factors such as emissions from lignite or the burning of petro-feedstocks, and from uncertainties about the rate of flaring at fossil fuel production sites.

It is important to realize that these uncertainties are compounded by the fact that consumption of fossil fuels entails additional emissions when the whole fuel cycle is taken into account. Additional energy use and associated emissions arise from production, transportation and storage activities. The emissions of CO_2 from fossil fuel combustion cause about 50% of the greenhouse effect. The remaining 25% referred to above is caused by non-CO_2 emissions. Methane losses are associated with the mining of coal and oil and the distribution of natural gas. Nitrous oxide is emitted by combustion of fossil fuels. The rate of this emission may be smaller than suggested some years ago because of a recently found sampling artefact, but may still be considerable in specific circumstances. Nitrogen oxides and non-methane hydrocarbons are precursors to ozone in the boundary layer. Apart from hydrocarbons incomplete combustion processes generate carbon monoxide, which competes with methane for OH-radicals, thus prolonging the atmospheric lifetime of methane. CO together with NO_x also produces the greenhouse gas O_3 in the troposphere. Furthermore the emission of aerosols plays a still uncertain role in climate. CFC use is also related to energy use (cooling). The emissions of non-carbon dioxide trace gases can be combined with CO_2 emissions using the concept of CO_2 equivalents. It is beyond the scope of this introduction to elaborate in detail on the influence of the indirect and equivalent emissions on the relative importance of different fuel types. For the purpose of this book it is sufficient to note that the addition of indirect and equivalent emissions does increase the contribution of the fossil fuel sector to the greenhouse effect, but in general does not alter the relative differences between the fossil fuels significantly.

The options

This book contains a collection of studies on the various options to control CO_2 emissions from the energy sector. The contributions are from specialists in many different relevant fields and show the existing expertise in the Netherlands. The combined views on the feasibility of controlling CO_2 emissions give an insight into present thinking within the scientific community in the Netherlands about possible responses to the greenhouse effect. During the last couple of years a change has been noticed in the discussion on climate change. The seriousness and urgency of the problem has been more widely acknowledged and at the same time the fundamental pessimism about the feasibility of response action has slightly decreased.

The contributions give a wide variety of technologically feasible options to limit CO_2 emissions and concentrations in the following fields:
- energy conservation
- change in fuel mix taking into account the estimated supplies (of for instance gas and uranium) in the world
- renewables (biomass, hydro- and windpower, solar energy)
- the creation of sinks by reforestation and biomass
- the removal of CO_2 from combustion gases followed by storage underground or in the ocean
- changing of the energy infrastructure (electricity, hydrogen).

The book also contains the following three integral studies:
- Long term scenarios up to the year 2100 which show the development of global emissions leading to a doubling of the CO_2 (equivalent) concentration in 2030, 2060 and 2090 respectively. A scenario which would give a sustainable situation has also been developed. Finally the so-called delayed response scenarios show the effect of an early or a postponed reaction to the threat posed by the greenhouse effect.
- A description of a feasible energy policy for the Netherlands which would give a cost-effective reduction of the national CO_2 emissions by 50% in the year 2020.
- A study of feasible energy options to determine whether CO_2 emissions can be stabilized in the Netherlands in the long term (in 2015).

The time horizon of the penetration of the different technologies varies, depending not only on technological development but also on future fuel prices and environmental policies, and last but not least on the social acceptation.

Long term global scenarios

Swart and Rotmans set the stage in chapter 3 by asserting that in order to limit climate risk sufficiently with the aim of enabling a sustainable economic and ecological development, a worldwide reduction of CO_2-emissions by more than 50% is necessary. Taking into account the ambitions and possibilities in the developing world this might necessitate a reduction by more than 80% in the industrialized countries. The authors base their argument on calculations made with a long term policy model that links the causes of climate change to the effects. The model facilitates an integrated evaluation of different international strategies enabling a balanced review of response actions in different sectors and aimed at different trace gases.

Such reduction percentages cannot be achieved by one solution only. A stepwise adaptation of the energy system will be necessary, starting with presently available and economic feasible options. The most immediate solution in terms of both economic and technological feasibility is energy conservation.

Energy conservation

Bosma of the Dutch Foundation for Energy Conservation (SVEN) argues that the most important cause for the waste of energy in industrialized countries is the disconnection between the functions performed and the associated energy consumption. Low prices and high levels of sophistication of equipment increase the lack of interest in energy conservation among consumers and industrial employees. He suggests that about 15% of the energy can be conserved by changes in behaviour and organization, which require no or negligible investments. By economic feasible investments for techical improvements another 15% can be saved. These figures are primarily based on the energy audits in the industrial sector which is co-ordinated by SVEN. Energy audits in industries can play an important role in the identification and stimulation of energy conservation in industry.

Becht and Van Soest of the Netherlands Centre for Energy Conservation and Environmental Technology emphasize the importance of the disengagement between services and energy use as a major cause of energy wasting. They also distinguish between the efficiency of equipment and the structure of the economy as two factors that influence the effect of energy conservation on the energy intensity of our economies. While generally the efficiency of equipment increases, this gain is often outweighed by structural changes associated with economic development (e.g. bigger cars and more electric household appliances). They analyse the possibilities for end use energy conservation in industry, transportation, households, services and agriculture. Similar to Bosma in the preceeding chapter they conclude that energy intensity can be decreased by about 2% annually by efficiency improvements in the coming two decades. Struc-

tural changes in the industrial and transportation sectors
may add another 1%. Taking into account the expected economic growth over this period, no more than a stabilization
of energy consumption could be achieved. As long as prices
of fossil fuels do not include environmental costs future
risks and long term scarcities the authors do not expect
any movement towards sustainable development. A combination
of price policies will be unavoidable. Revenues from carbon
based taxes can be used as subsidies for investments in
energy conservation and sustainable energy. Such price policies should be complemented by efficiency standards, task
setting of utilities and an extensive technology development and transfer programme.

Diepstraten and Van Gool approach the problem from the
viewpoint of maximizing exergy efficiency. Exergy is the
thermodynamic quality of energy and material resources,
including not only the amount of joules, but also what the
joules are used for (heat, traction, etc.). Objective of
this maximization is the matching of demand and supply. The
sectors households and industry have the best potential to
increase energy productivity and consequently decrease carbon dioxide emissions. The authors suggest that next to the
obvious fuel shift in electric power production the use of
electric heat pumps in dwellings and buildings, restructuring of industry by better heat management, electric vehicles for urban transportation, physical removal of carbon
dioxide from stack gases and chemical reuse of carbon dioxide are promising options. This last option may involve
the production of durable chemicals or carbon based fuels
like methanol, thus delaying the final emission of carbon
dioxide. Taking into account the lead times for the introduction of these options, the authors find that in The Netherlands a stabilization of carbon dioxide emissions in
2015 at present levels would not be possible because of the
effect of increasing economic growth. Most options would
imply a shift towards a more electric society.

Renewables

Although with energy conservation we might go a long way
towards reducing emissions of carbon dioxide, it will be
far from sufficient to achieve the reduction percentages
mentioned above. As cited by Lysen in his chapter on renewables, the World Commission on Environment and Development
'believes that every effort should be made to develop the
potential for renewable energy, which should form the foundation of the global energy structure during the 21st century'. From this perspective all other options to be discussed later in this book should be considered as transition stages. Lysen evaluates the feasibility of the expan-

sion of the use of biomass, hydropower and wind power from a global perspective. With respect to biomass he argues that theoretically the existing forested area would only have to be increased by 7% to balance fossil carbon emissions. This reafforestation should be accompanied by the gradual replacement of fossil fuels by biomass or other renewables in order to make it more than a temporary solution. An important suggestion is that reforestation at the higher latitudes is as important and attractive as the reversal of the deforestation process in the tropics. Costs of wood for energy production in Europe would be in the same order of magnitude as present fossil fuel prices. Next to biomass also hydropower can play an increasing role in the future world energy supply. In many places this resource can be tapped without ecologically disruptive effects. For The Netherlands wind and (maybe surprisingly) especially solar energy are more interesting than hydro power. The author argues that prices move slowly in a more competitive direction. In terms of CO_2-reduction potential per unit of land area solar electric technologies are more productive than wind or biomass, but these technologies still have a cost disadvantage. Low costs and absence of important side-effects make biomass an interesting be it partial solution.

While the previous author looked at the biomass energy potential from the theoretical viewpoint, Wiersum and Ketner of the Wageningen Agricultural University have their doubts about the practicability of this option. They stress the importance of a better understanding of the biotic part of the carbon cycle for reliable calculations of the quantitative potential of carbon fixation by reforestation. According to some scientists the carbon dioxide fertilization effect and the charcoal formation during burning of vegetation might outweigh the emissions by burning, causing the biosphere to be a sink for carbon rather than a source.

Stopping deforestation in the tropics and abating forest dieback in the temperate zones would limit carbon losses to the atmosphere and should be pursued. Also forest plantations for carbon sequestering is an attractive option. The actual potential of such schemes very much depends on the management, including rotation times. The use of wood as a substitute for fossil fuels would lead theoretically to an energy system based on a closed carbon cycle. Since wood is a bulky product with low energy content per weight unit plantations should be located in the vicinity of the users. The authors also emphasise that there are other, usually far more pressing needs for reforestation, such as the rapidly increasing demand for fuel wood in rural areas and timber and the need to avoid erosion. Therefore at most carbon fixation by forest plantations will be

one additional objective that can be taken into account
when designing optimal afforestation schemes in specific
situations. Plantations for environmental conservation purposes and industrial wood plantations offer the best possibilities for this combination. These authors assume the
practical biomass fixation potential of forest plantations
mcuh lower than the previous author. Assuming similar
amounts of plantations as Lysen the long rotation plantations may fix only about 0.5 Gt C annually. Additionally
local short rotation fuel wood plantations might add another 0.1 GtC/year and again a similar amount can be added
supposing that fuel wood can partially replace fossil
fuels. Since optimal production would take place in soils
that are also suitable for other (agricultural) practices,
this option is interesting for the lands taken out of production in the temperate zones. In the tropics carbon sequestering can at most be regarded as a welcome by-product
of reforestation projects set up primarily for other purposes.

From the renewable energy sources solar is generally
claimed to offer the best opportunities for the widest application. Turkenburg of Utrecht University examines the
prospects for photovoltaic conversion. He argues that at
the end of the nineties PV can capture a significant share
of small scale stand alone applications and peak load supply. This development is likely to be followed by much wi-
der application, including base load supply, triggered off
by decreasing prices within two decades. This may be shifted forward with about ten years when social and environmental costs of energy systems would be taken into account.
The environmental problems associated with solar energy,
including the use of potentially harmfull materials that
have to be controlled, are small compared to those of fossil fuels, certainly as far as the greenhouse effect is
concerned.

The estimated supplies

In this book as well as in the public debate about possible
response to environmental threats from the energy sector
the expansion of the use of natural gas and nuclear energy
is proposed as having more potential at the short term than
renewables. Since it is important that the direction in
which the energy system in view of environmental problems
will be modified can be sustained for a long time, in chapter 14 De Vries of Groningen University discusses the long
term availability of gas and uranium. The different dynamic
characteristics of reserves as a function of eventual resources, associated grades, costs and energy ratios are
analysed systematically in detail. The author concludes

that present estimates of proven and unproven methane do not indicate any major resource constraint when methane would capture 50 to 60% of the world energy supply over the next half century, provided that end use would be stabilized. If cumulated methane requirements would necessitate the exploitation of yet speculative reserves at probably higher costs and lower net energy gain the carbon dioxide substitution potential of natural gas would be lowered. High penetration of natural gas might be thwarted by the capital intensity of the necessary infrastructure and by political considerations because of the concentrated presence of reserves in a limited number of countries rather than by physical resource constraints. As to nuclear energy, a penetration of 10 to 20% of the world energy demand at present levels can be met by the reasonably assured and estimated additional reserves of uranium for a prolonged period. The carbon dioxide gain by application of nuclear energy might decrease rapidly when lower grade ores have to be used. The combined potential CO_2 reduction is estimated to be 15 - 20% over the next half century.

Removal and disposal

A longer term solution might be the recovery of carbon dioxide from combustion point sources. Although in the recent past in many countries the prospects of this technology have been dim, Hendriks, Blok and Turkenburg of Utrecht University find this option rather promising. Next to the often mentioned chemical absorption at conventional power plants they suggest a new approach making use of a shift reaction and a physical absorption process in combination with an integrated gasification combined cycle plant. Only moderate losses of conversion efficiency and electricity price increases of about 25% would result. This option would fit nicely in the end-of-pipe approach that has often been applied to combat environmental problems. Taking into account the enormous political and economic difficulties involved in preventing the abundant coal reserves from being exploited, it seems premature to shelve this option. The associated price increase would stimulate the introduction of other solutions more in line with the concept of sustainable development, like renewables and energy conservation. Nevertheless, it should be kept in mind that this option can only provide a partial solution, since probably it can only be applied to large combustion sources like power plants. Furthermore, as any end-of-pipe solution carbon dioxide recovery produces waste. Next to the small waste streams of potentially hazardous absorbens and catalysts the bulk will be formed by carbon dioxide in liquid or gaseous form that has to be removed. The authors do not include technical measures and associated costs of the final removal.

The capacity of salt domes in The Netherlands appears to be too small for storage of sufficient amounts of carbon dioxide over sufficiently long periods. Alternatively, usage of recovered carbon dioxide for instance for enhanced oil recovery has a certain potential, but provisional analyses indicate that these cannot make a significant contribution in The Netherlands to the problem of removing all CO_2 from power generation.

Another solution might be the storage in the deep ocean. But what are the physical and ecological consequences of such a solution? De Baar and Stoll of the Netherlands Institute of Ocean Sciences (NIOZ) argue that theoretically the oceans have an enormous capacity to store carbon. They stress the uncertainty with respect to the actual process removing carbon dioxide from the atmposphere. If the physical processes would predominate, the increasing carbon dioxide concentrations in the atmosphere would decrease the inorganic buffering capacity of the ocean, implying a dangerous feedback. If the biological mechanisms would be more important, the ocean uptake might develop more favourably. The assessment of both biological and chemical uptake are focal points of the Joint Global Ocean Flux Study in which NIOZ participates. The authors also refer to the uncertain ecological impact of concentrated injections with consequent increasing acidity. The major difficulty lies in the different time scales of rapid human interference with the carbon cycle and slow natural adjustment to this interference. An injection of CO_2 into the deep ocean would only be partially buffered by the dissolution of calcite because of the time necessary to reach equilibrium. Dependent of the place of injection the excess carbon dioxide will re-enter the atmosphere within hundreds of years. Therefore, apart from being expensive, partial and ecologically potentailly disadvantageous, deep sea injection can only be a temporary remedy in a transition period towards a CO_2-free energy system.

Van der Harst and van Nieuwland of the Dutch Oil and gas Company (NAM) review the options of CO_2-storage in empty gas fields. Next to the main Dutch reserve, the large Groningen field, recoverable gas reserves are distributed over a large number of small onshore and offshore fields. Carbon dioxide injection can only take place after definite abandonment of a field, excluding this possibility for the Groningen field, which is envisaged to be depleted around 2050. Nevertheless the authors calculate that theoretically the carbon dioxide production of the present power generation sector could be stored in the small onshore fields for some decades from the turn of the century. Costs would be about 10% of the costs of the removal of carbon dioxide from the stack gases and dehydration as discussed in chapter 10. Storage in offshore fields would be unattractive from technical and economic perspective. For large gas fields ultimate gas recovery by CO_2 injection, comparable to enhanced oil recovery, could be possible and might prolong the productivity of the fields. Since so far the major known gas reserves are located at a limited number of locations usually not close to the combustion sites, the potential of storage of CO_2 in empty gas fields is limited from a global perspective, but may be interesting in specific regions.

Electricity and hydrogen

A large portion of the response options dealt with in the preceeding paragraphs depend on the increase of electricity in end use energy consumption. Not only the removal of carbon dioxide but also the introduction of other types of low carbon energy sources might be facilitated by an increased use of electricity: a move towards an all-electric society. Van Wunnik of the Electric Utilities Research Institute (KEMA) evaluates the possibilities of such a shift for The Netherlands, concentrating on housing, traffic and industry. He argues that such a shift is likely even without the greenhouse problem. However, the effect of such a change on the emissions of carbon dioxide depends fully on the technological choices made during the expansion of the electricity sector. Even in the medium term nuclear and renewable energy and carbon dioxide recovery are unlikely to play a significant role in The Netherlands electricity sector. Increasing electricity use will increase these emissions as long as fossil sources serve as the major fuels in electricity generating plants. This effect can be counteracted to a certain extent by the introduction of more combined heat and power installations and the expansion of the utilization of efficient gas technology.

While the previous chapter dealt with the short and medium term outlook for electricity expansion, Muis and Slob of Consultants on Energy and the Environment (CEA) evaluate the long term potential of hydrogen as a secondary fuel. Fossil fuels have high energy densities, can easily be transported and stored, but have major environmental disadvantages, are exhaustible and not distributed evenly over the world. With renewables the situation is quite the opposite. The use of hydrogen as a secondary fuel produced with renewable energy sources combines the advantages of both types of fuel without the shortcomings. Although technologies necessary for the implementation of this attractive concept are demonstrated and available, the costs still prevent penetration. The authors suggest fossil fuels can be used for hydrogen production in a transition phase, taking into account that in the past it has taken 40 to 50 years for a new energy carrier to capture 10% of the market share. In terms of the greenhouse effect a major disadvantage of the system is that its high capital intensity and high level of technological sophistication will limit the applicability in developing countries. Nevertheless it is argued that the present price would equal the price of the present fossil fuels when the social and environmental costs would be fully taken into account. Furthermore many old energy systems were not replaced by new ones because they were cheaper, but because they were better. Most promising applications for the short term are the (air) transportation sector and the use of fuel cells for both heat and electricity generation. They calculate that an energy system in the Netherlands based for 20% on electricity and for 80% on hydrogen would cost about 8 billion guilders annually.

Optimal response

What would the combined potential of all these options be for an industrialized country like The Netherlands?
In the last chapter Kram and Okken present an integrated analysis based on calculations with the MARKAL-model. They consider the possibilities of a 50 % carbon dioxide emission reduction in 2020 as compared to 1985 for two scenarios: a nuclear supply-oriented pathway and a demand oriented gas strategy. They find that in both cases the target reduction can be achieved, be it with considerable costs, comparable to those planned to combat acidification. Also the resulting energy systems differ widely. The nuclear scenario clearly enters the road toward an all-electric society. Notwithstanding these differences, some options are common for both scenarios and it is recommended to stimulate the introduction of these technologies. They include most of the options discussed in this book, including energy conservation, materials recycling, renewables and high efficiency gas conversion technologies. The approach followed is an analysis of options within a consistent framework of fuel prices, environmental constraints, demographic, economic, industrial and technological developments. A minimum carbon tax on fossil fuels is calculated that would be required to identify and implement cost-effective CO_2-reduction strategies.

In a brief concluding paper van Gool expands on his views of the policy potential of the options. The implementation of the options would need a world wide continuous effort over several decades. Recognition of environmental problems in the world generally being low and governments seldom being in power for more than 4 to 5 years, the greenhouse problem can not be solved without changing the political and social structures of our societies. Invisibility of the effects of drastics policy measures and the unlikelihood of governments to transfer a part of their sovereignty to a higher entity will make the implementation of the necessary mechanisms a difficult process.

3. OUTLOOK

Notwithstanding the political and economic barriers that have to be overcome the drift of most papers is definitively positive. Yes, the greenhouse effect can be slowed. Not one option alone will be sufficient to reduce the emissions to a significant degree, but according to the different authors technologically it is possible to reduce the emissions of carbon dioxide with tens of percentage points. The options vary in time of availability, costs and effectiveness.

From the different papers together no clear picture of a low-CO_2 future emerges immediately. Many more options are available to limit emissions than was generally assumed some years ago, but there is no easy way out. The difference with other air pollution problems, like acid rain, lies in the enormous quantities of emitted gases and the direct linkage with energy consumption and lifestyle i.e. a person using one tank of automobile gasoline emits twice his own weight in carbon dioxide into the atmosphere.

The most effective option that can be applied immediately, is energy conservation in the broad sense, including not only technological efficiency improvements but also institutional, organizational and behavioural changes. With a maximum commitment and effort it should be possible to reduce increases in energy use and associated emissions of greenhouse gases without a disruption of the economic development.

To be able to go further than stabilizing emissions, fuel switches to other 'hard' energy sources, notably natural gas and nuclear energy, come into play in the short run. It is expected that a combination of these options will reduce emissions by no more than about 20%. One reason for this is that the dwindling resources of uranium and gas will lead to increase costs and energy use in production. Secondly natural gas still produces carbon dioxide, be it at lesser rates than oil and coal and finally the penetration of nuclear energy will be thwarted by its limited applicability.

With a somewhat longer time perspective the removal and disposal of carbon dioxide can play a role. Recent insights suggest that energy price increases may become acceptable. Disposal could take place in depleted natural gas fields, which limits the geographic applicability of this option. Disposal in the ocean is a questionable and only temporary solution. Only suitable for large point sources in a limited number of regions, this option might add another 5 to 10% CO_2 reduction.

Biomass has a considerable potential. Forest plantations do not only serve as a sink for carbon dioxide, they can also produce wood as a replacement for fossil fuels. Because of the practical limitations in developing countries this option should also be explored more carefully for the temperate zones. Estimates of the contribution of this option to the reduction of net carbon dioxide emissions differ from 5 to 100%. A prudent estimate would be about 10%.

The only way a real long term sustainable energy future with low CO_2 emissions can be achieved appears to be a shift towards renewable energy, in which solar energy has to play a central role. The rapid development of photovoltaics is an important step in this direction. Other - "nonrenewable" - options (gas, nuclear energy, carbon dioxide removal) are important in this respect for a transition phase of at least several decades. Tied up with the introduction of renewable energy sources is the further development of benign energy carriers: increasing use of electricity to generate power and development of hydrogen technology for heating and transportation. (An energy system based for 20% on electricity and for 80% on hydrogen would cost the Netherlands about 8 billion guilders annually.) To have the desired environmental benefits the electricity and hydrogen should be produced with renewables at optimal and central locations.
If the renewables indeed represent the most promising option the immediate incorporation into our energy policies and planning is vital.

IMAGE: A TOOL FOR LONG TERM GLOBAL GREENHOUSE POLICY
ANALYSIS
An Integrated Model for the Assessment of the Greenhouse
Effect

R.J. Swart and J. Rotmans
National Institute of Public Health
and Environmental Protection (RIVM)
P.O. Box 1,
3720 BA Bilthoven, The Netherlands

ABSTRACT. The rapid recognition of the projected climate
change as an important issue by politicians over the last
few years has brought about the necessity of a methodology
to assess different global policy options. In this chapter
an example of such a methodology is described: the Integrated Model for the Assessment of the Greenhouse Effect
(IMAGE). Next to enabling the evaluation of long term climate strategies the model was developed to increase awareness among different societal groups, to give an overview
of the complex problem to different specialists and to
identify gaps in knowledge. Some recent calculations made
within the framework of the Intergovernmental Panel on Climate Change are presented.

1. INTRODUCTION

Even faster than the depletion of the ozone layer the anticipated climate change has come to play a role in high
level policy debates. Since the problem is very complex by
nature a need has arisen among policy makers to dispose of
a tool that gives a clear and concise overview of the work-
ings of the greenhouse effect and the relevance of potential policy options. The Integrated Model for the Assessment of the Greenhouse Effect (IMAGE) was developed at the
National Institute for Public Health and Environmental Protection from 1986, at a stage, when primarily the recognition of climate change was intended to be its primary role.
Since the enhanced greenhouse effect is created by a multitude of causes and may result in a similar multitude of
effects, the model tries to capture these causes and ef-

fects in an integrated fashion. The main difficulty thwarting policy response to the greenhouse effect is that the causes are not only many, but are also the fundamental basis of our society: the present practices with respect to energy production and consumption, agriculture and industry. Dependent on the assumptions and definitions it can be generally said, that the agricultural sector including deforestation causes 25 % of the problem, while 75 % is caused by the energy and industry sectors. Within the energy sector transportation, power generation and other combustion processes play about an equal role (see figure 1).
The model was used as a demonstration tool for a large number of groups of politicians, scientists and the general public. In section 2 a brief description of the structure of the model at its present stage of development is given, followed in section 3 by calculation examples as made for the Intergovernmental Panel on Climate Change.

RELATIVE CONTRIBUTION TO CLIMATE CHANGE

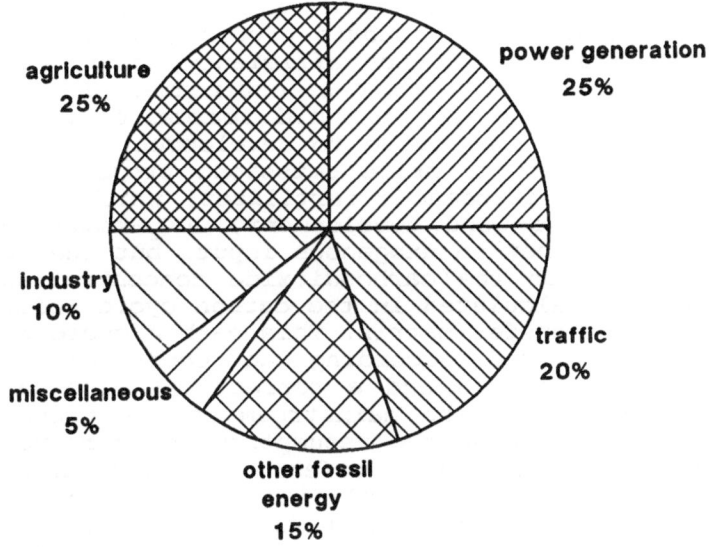

Figure 1

2. MODEL STRUCTURE

At its inception IMAGE was set up primarily as a tool for long term greenhouse policy analysis and demonstration sessions, running on a microcomputer[1]. The fast technological development of microcomputers enabled RIVM to expand the original simple parameterized model structure by a number of more detailed modules without interfering with the original purpose. Basically IMAGE tries to capture as much of the cause-effect relationship with respect to climate change as possible. The causes are considered at the global level while so far impacts beyond global mean temperature and sea level rise are included for the Netherlands only.

The structure presented in figure 2 shows a number of independent though interlinked modules, each describing a specific element of climate change.

The trace gases CO_2, CH_4, N_2O, CO, CFC11 and CFC12 are presently taken into account. Ozone and ozone depleting substances other than 11 and 12 are planned to be added in the near future. Different sources are distinghuished, grouped as nature, industry, agriculture and industry. At present the most detailed emission module is the energy module, for which the pc-version of the Edmonds and Reilly model is used[2]. Next to carbon dioxide emissions also methane emssions from coal mining and gas distribution and nitrous oxide emissions from combustion are related to the results of scenario calculations with this model in a simple way. The main parameters that are varied during scenario analysis are population, economic growth (labour productivity), end use and conversion efficiency, cost developments for non-fossil energy sources, environmental costs and carbon taxes. So far the agricultural sources of methane and nitrous oxide are scaled to developments in the relevant areas (e.g. nitrous oxide emissions proportional to the consumption of nitrogenous fertilizer, which again is a function of arable land development). CO_2-emissions from deforestation are not an exogenous input, but the impact of deforestation on the carbondioxide concentrations in the atmosphere is calculated in the carbon cycle module by way of a landuse transfer matrix. This carbon cycle module, which is simulated according to Goudriaan and Ketner[3], has an ocean module and a terrestrial biosphere module. In order to follow the mainstream of the present scientific opinion we relaxed the assumptions with respect to the CO_2-fertilization effect as referred to in Goudriaan and Ketner's paper. For CH_4 a simplified atmospheric chemistry is simulated in which the methane concentrations depend on the abundance of OH-radicals which again depends on the carbon monoxide concentrations[4]. Since a large fraction of the increase of the concentration of methane in the atmosphere is most probably caused by CO competing for OH-radicals, inefficient combustion in the energy sector also contributes to the greenhouse effect via this route. For CFCs a constant atmospheric lifetime is assumed. The delay time between production and emission is assumed to be different for different applications. Finally nitrous oxide concentrations are computed from emissions by taking into

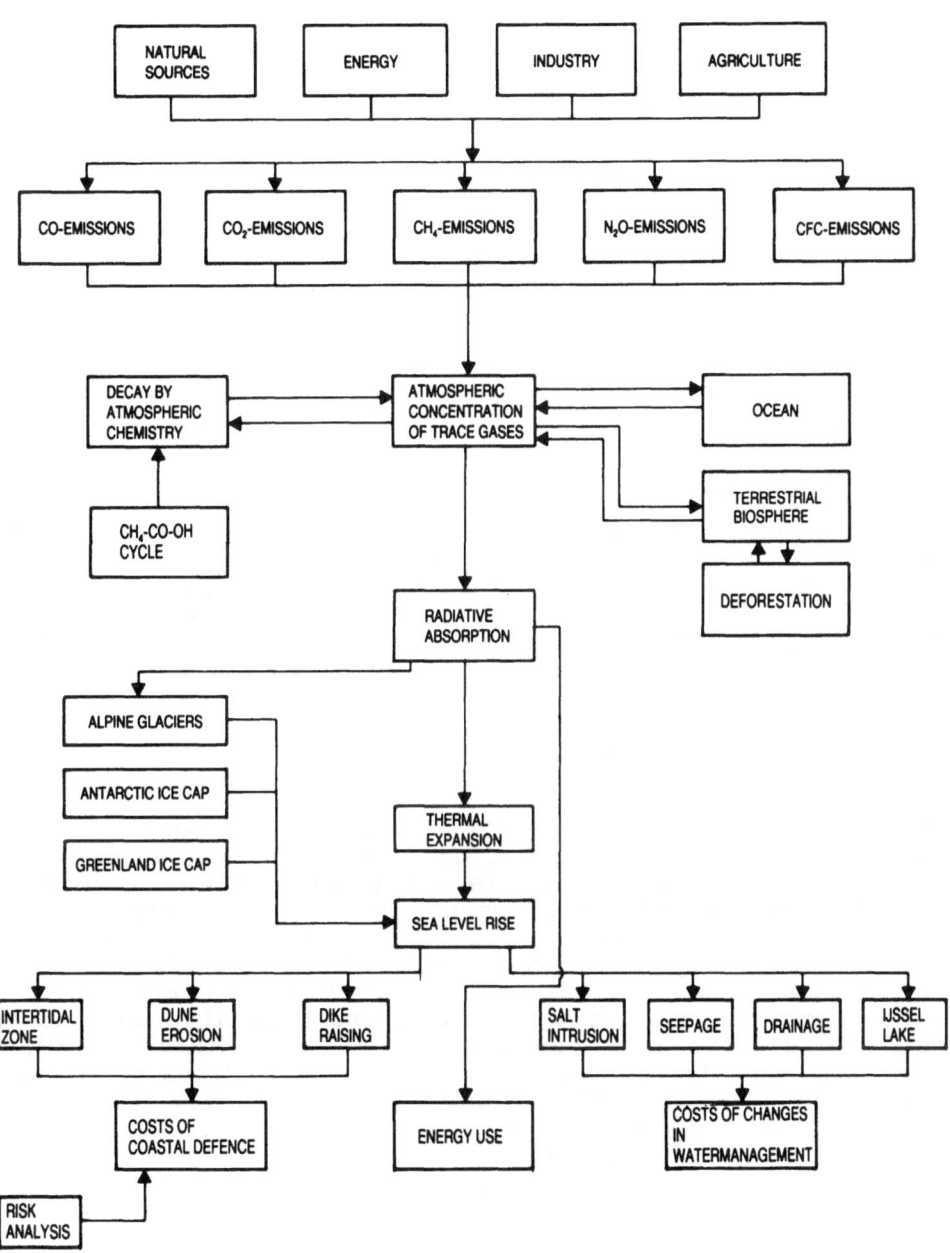

Figure 2

account a constant atmospheric lifetime.

The parameterized radiative convective module including different feedbacks is based on work by Wigley[5]. The
resulting temperature changes are the input for the sea
level module. In this model, based on findings by Oerlemans[6] the effects of global warming on the sea level are
determined by the thermal expansion of the ocean waters,
the changing ice caps and the observed natural sea level
rise. This global sea level rise is not yet regionalized.
Nevertheless, for different values of accelerated sea level
rise in the North Sea a number of consequences for the Netherlands' coast is modelled: the impacts on dikes, dunes
and intertidal areas. Finally for four elements of inland
water management the impact of sea level rise has been modelled: salt intrusion, seepage, drainage and the management of the IJssel Lake[7].

Recently IMAGE was compared to other similar models[8,9].
Results of model calculation with the same input data gave
a good match in terms of temperature effect. Generally the
emission (energy) modules are somewhat more detailed in the
other models, while IMAGE includes effect modules, which
the other models do not. Future work comprises the improvement of user friendliness, the performance of uncertainty
analyses, improvement and full intergration of the emission
modules and addition of other substances.

3. SCENARIO CALCULATIONS

The main application of IMAGE is the development of long
term scenarios (up to 2100). It has been used extensively
in the Netherlands over the last few years to evaluate different future worlds with respect to climate change. Next
to demonstration sessions for different groups the model
was used for different Dutch environmental studies, such as
the preparatory document[10] for the National Environment
Plan. Recently IMAGE entered the international arena. Some
scenario results from calculations made within the framework of the Intergovernmental Panel on Climate Change
(IPCC) will be briefly discussed here. Early 1989 the
Steering Committee of the Response Strategies Working Group
of IPCC requested the USA and the Netherlands to prepare a
document on three emission scenarios for use in different
IPCC-working groups. Initially these scenarios had to be
designed in such a way that they would lead to a doubling
of CO_2-equivalent concentrations in the years 2030, 2060
and 2090 subsequently. In the last scenario the concentrations would stabilize afterwards. The Dutch contribution to
the draft report[11] that was prepared jointly with EPA con-
sisted next to the determination of the conceptual ap-

proach, of the performance of a number of IMAGE calculations. First scenarios made with EPAs Atmospheric Stabilization Framework were evaluated and secondly a number of
additional analyses were made. Since for the purpose of
IPCC the consequences of policies in the decades to come
are of crucial importance some runs were performed simula-
ting delayed response actions.

Low climate risk scenario

The Netherlands and some other countries were not satisfied
with the range of scenarios selected, primarily since even
the lowest scenario would lead to temperature increases
that are unprecedented in the last 100,000s of years. Not
considering the feasibility of pursuing such a scenario
(which is extremely low) some calculations were made to
assess tolerable emissions which would limit climate change
risks sufficiently. Such a low scenario is important for
the assesssment of the reaction of the climate system by
scientists and is the first step towards painting a picture
of a sustainable world. It is not yet included in the draft
report referred to above. In figure 3 the resulting CO_2-
scenarios are shown. It appears that the highest IPCC-sce-
nario would allow an increase of CO_2-emissions by a factor
of 4 in 2100, while in order to follow the lowest pathway
these emissions can only grow by about 40 % towards the mid
21st century as compared to 1985. For comparison we drafted
the expected carbon dioxide emissions associated with the
most recent IEA World Energy Outlook[12]. If world energy use
and fuel mix would follow this projected path, the emissions of CO_2 and other trace gases would exceed even the
highest scenario as taken into account by IPCC. To put the
scenarios into an even more pessimistic perspective: in
the lower scenarios it has been assumed that the emissions
of the other trace gases can be limited within the same
time scedule as CO_2 be it at different levels. For regulated CFCs this might not cause problems, but for gases with
as yet not fully quantified sources like methane, nitrous
oxide and carbon monoxide this tends to be an optimistic
assumption. When these limitations could not be achieved,
the necessary control of carbon dioxide should even be
stricter than envisaged in the presented IPCC-scenarios in
order to reach the same goals. In terms of geographical
distribution the lowest scenario allows for an increasing
emissions in the developing countries and more or less stabilizing emissions in the industrialized countries. In the
Netherlands National Environment Plan this stabilization
(to be achieved in 2000 at 1989 levels) has been included
as a provisional goal for the emissions of carbon dioxide.
The global low climate risk scenario in figure 3 shows
a gradual decrease of global CO2-emissions which is more or

24

IPCC AND LOW CLIMATE RISK SCENARIO

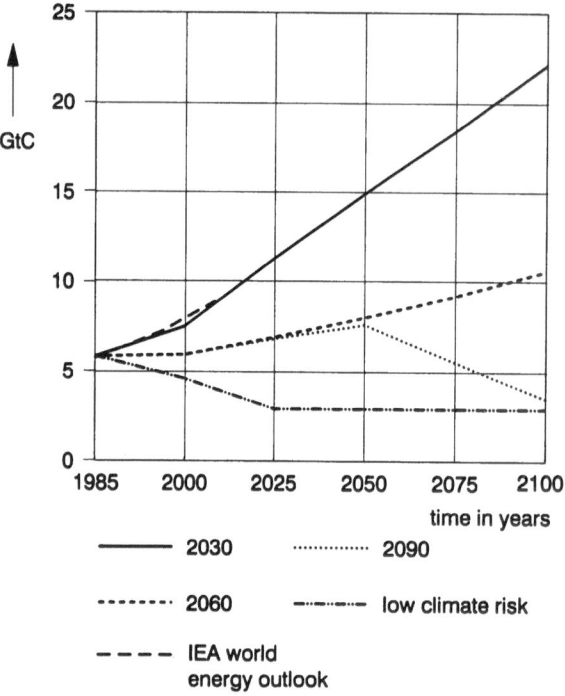

Figure 3

less consistent with the Toronto-recommendations if
applied world wide: 20 % reduction in 2005 and 50 % reduction by the half of the next century. In chapter 16 is calculated that such reductions are possible in the Netherlands after 2020. Worldwide this pathway could only be
achieved by a massive and rapid change of the world energy
system towards maximum energy efficiency and the application of CO_2-free technology. Although such a change might
lead to continuing or even accelerated economic growth for
some groups in some countries, there is a risk in major
parts of the world these emission reductions might only be
achieved by limitations to quantitative economic growth
which are unacceptable from the viewpoint of the present
day national ambitions.

Delayed response

In figure 4 the emissions of CO_2 for different delayed response runs are depicted. It has been (optimistically) as-

DELAYED RESPONSE ANALYSES

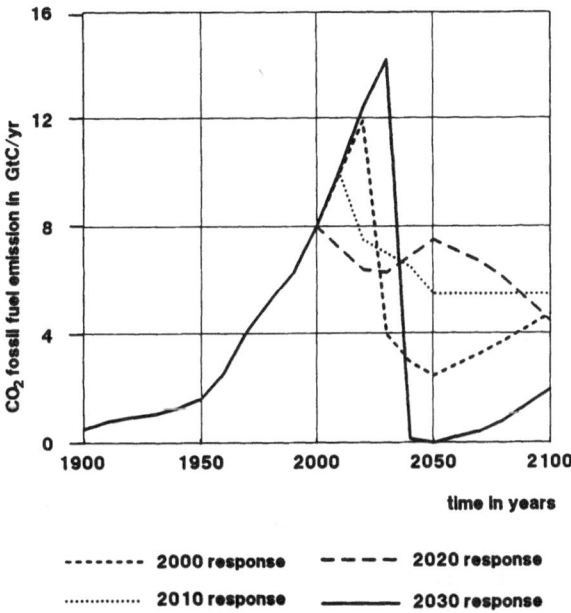

Figure 4

sumed that emissions of trace gases would follow the emissions of the 2030 doubling case when no action would be
taken to limit greenhouse gas emissions. Furthermore it is
assumed that when the international decision would be taken
and followed up to start controlling climate change, the
policy target would be stabilization of concentrations in
2090 at double CO_2-equivalent levels (the lowest IPCC-sce-
nario). Emissions of non-CO_2-trace gases were assumed to be
deflected towards the values associated with the 2090 scenario and consequently the allowable emissons of CO_2 were
determined. It appears that waiting until 2000 would implicate only a slight necessary decrease of emissions. Waiting
10 years more would necessitate gradual emission reductions
of about 40%. Again 10 years later in 2020 almost 70% emission reduction would have to be achieved in less than 10
years in order to reach the target. As can be expected because of the choice of base line emissions (leading to 2030
doubling) delaying response until 2030 would render a complete and swift phase-out of carbondioxide emissions from
fossil fuel combustion necessary.

4. DISCUSSION

Over the last few years the usefullness of an integrated methodology to evaluate different futures with respect to climate change has been proven extensively. IMAGE gives clear long term relationships between causes and effects of the enhanced greenhouse effect, based on the latest scientific evidence, but without requiring detailed scientific knowledge from the user. The Edmonds and Reilly model having 25 year time steps, IMAGE is less suitable to assess details of the effects, feasibility and costs of different policy options in the short run. Attention to this drawback will be given over the following years in collaboration with other groups. Nevertheless the results so far indicate that really drastic steps have to be taken, primarily in the energy sector, not to let climate change get out of hand.

The projections of IEA[1] 1 show, that if the greenhouse effect is not taken into account in energy policy planning, emissions of greenhouse gases will show accelerated growth when the present world economic growth continues based on the same energy system as before. Abundant reserves of cheap coal will be used unless technological and financial mechanisms are developed to push energy supply development in a more benevolent direction. The papers in this book show that technologically the Netherlands and the world can go a long way towards limiting emissions. Dependent on the economic and technological characteristics and preferences future fuel mixes will vary per country. Different options will become attractive at different times, but in all cases increased energy conservation in the broad sense should be the first to implement because of its favourable economic potential and present availability. Economic and institutional barriers to the introduction of the technologies discussed in this book have to be pulled down as quickly as possible because of the long lead times for full penetration and the urgency of the problem. According to our analysis the emissions of CO_2 in the industrialized countries should be decreased with at leats 80 % over the coming 50 years in order to prevent serious risks for the world community. Delays in policy formulation will be followed by long lead times for introduction of measures increasing the amount of climate change 'in the bank' because of the slow response of the physical climate system. Therefore rapid response action is warranted.

5. LITERATURE

1.   Rotmans, J., de Boois, H. and Swart, R.J.: 'IMAGE: An Integrated Model for the Assessment of the Greenhouse Effect', RIVM report nr. 758471009, Bilthoven, 1989
2.   Edmonds, J.A. and Reilly, J.M.: 'The Long-term Global Energy-CO_2 Model: PC-version A84PC', Carbon Dioxide Information Center, Oak Ridge, 1986
3.   Goudriaan, J. and Ketner, P.: 'A Simulation Study for the Global Carbon Cycle including Man's Impact on the Biosphere', Climatic Change 6, 167-192
4.   Rotmans, J. and Eggink, E.: 'Methane as a Greenhouse Gas: a Simulation Model of the Atmospheric Chemistry of the CH_4-CO-OH-cycle', RIVM report nr. 758471002, Bilthoven, 1988
5.   Wigley, T.M.L.: 'Relative Contributions of different Trace Gases to the Greenhouse Effect', Climate Monitor 16, no. 1
6.   Oerlemans, J.: 'Greenhouse Warming and Chnages in Sea Level', paper presented at workshop 'Developing Policies for Responding to future Climatic Change', Villach, 1987
7.   Elzen, M.G.J. den and Rotmans. J.: 'Simulation Model for a Number of Socio-Economic Consequences of the Greenhouse Effect for the Netherlands' (in Dutch), RIVM report nr. 758471008, Bilthoven, 1988
8.   Mintzer, I.: 'A Matter of Degrees: The Potential for Controlling the Greenhouse Effect', World Resources Institute, Washington, 1987
9.   EPA: 'Policy Options for Stabilizing Global Climate', draft report to Congress, Washington, 1989
10.   Langeweg, F. (ed.): 'Concern for Tomorrow', RIVM, Bilthoven, 1989
11.   IPCC: 'Emissions Scenarios of the Response Strategies Working Group of the Intergovernmental Panel on Climate Change', Expert Group on Emission Scenarios, Bilthoven, 1989
12.   IEA: 'World Energy Outlook', draft paper, IEA, Paris, 1989

HOW TO DECREASE THE CO_2 EMISSIONS WHILE SAVING MONEY

ir. J. Bosma
SVEN, Voorlichting energiebesparing
Postbus 503
7300 AM APELDOORN
The Netherlands

CO_2 emission is created by using fossil fuels for energy generation. In our society the efficient use of energy is still not on the level that can be attained by using well known and existing techniques.

The sophisticated modern equipment now on the market and the low prices of energy have created a use pattern that can be characterised by:
- the user's indifference to the real energy consumption of the equipment he uses in his daily work.
- a non existing direct connection between the function of the equipment and energy consumption.
- the use of sophisticated equipment by users who do not have any technical knowledge of the equipment.

The user has a bare minimum of instructions on how to use his equipment, if something does not perform as he thinks it should, he is unable to detect the malfunctioning and so the help of more specialised people has to be called in. This creates a distance between the user and his equipment, he is not interested in energy efficiency for his work. And so a waste-market for energy use is created.

In the private households this is illustrated by the following examples:
- in the case of a coal fired stove, the labour needed for heating was directly experienced. If the weather was good one knew this immediately by having less bins of coals to furnish.

  When, however, the heating system is gas fired, the connection between heat and energy is less visible and when the weather improves the room-temperature is controlled not by turning off the heating, but by opening the windows.

Furthermore, the modern heating system has its own control system. The bills from the gas company are automatically paid and are not used as a signal. So when the temperature drops in mid-summer in the small hours, the heating system is started and will supply heat nobody called for.

Sometimes the installed power is used as a sales argument. Some time ago vacuum cleaners had 200 Watts motors, nowadays the power installed is more than a kilowatt. One may wonder if this is needed.
The same trend is shown in the transport sector. It remains an open question whether the now installed power in private cars is really needed for day to day work.
Traffic congestion lowers the average speed. The only profit we have in our modern cars is that we can reach that average speed in a shorter time.
With these straightforward examples I intend to indicate that, although we do have access to modern and sophisticated and energy efficient equipment, there are a number of barriers in using this equipment properly. This results in energy use that is (highly) inefficient. The question arises how much energy is wasted.

SVEN the Netherlands foundation for energy conservation, it was founded in 1976. Sven employs about 50 staff members.
The task of SVEN is to promote energy conservation. This is done via leaflets, brochures, seminars, exhibitions and specific action. One of the more specific projects of Sven is the so-called busproject. With this project one day energy audits are carried out in industry and in organizations.
The promotion of energy conservation has been quite successful. Hard facts about the effect of promotion are difficult to obtain for when an individual is successful in energy conservation it is the result of his own efforts; the Sven promotion can only trigger these efforts. Nevertheless, as an example of results Sven achieved during its existence I shall mention the following figures: in 1986, ten years after SVEN's establishment, we conducted an inquiry that yielded the following results.
In the period 1976-1986 some 8,000 energy audits took place. Of all the possible and proposed options for energy conservation 50% had been implemented. The total investments for this implementation were 925 million guilders. The revenues of these implementations during the same period were 950 million guilders. Another inquiry shows that promotional activities of Sven will result in 4-25% more options to be implemented. The Sven activities created a profit to society of 40-250 million guilders in the first ten years of the foundation's existence.

One of the tasks of Sven is the execution of the so-called grant scheme for energy audits.

Energy audits are executed by independent consultants. If they follow the rules laid down in the law, a fee of 40% of the costs is paid. One of the rules is that a concise copy of the audit results is presented to Sven. So far about 14,000 of these audits have been made in The Netherlands.
The audit results presented to Sven must contain:
- energy balance
- a list of conservation options, the expected results of the option and the costs
- a list of options to be considered for implementation.
The energy figures resulting from these audits are related to the final energy use.

Energy use and energy conservation is determined by three categories: organization, human behaviour and technical solutions. We classify the options in these categories. Options regarding organization and human behaviour are options that need low investments or non-investments.
The implementation for technical solutions call for investments. Investments are made if the economic feasibility is favourable.
The 'waste market' of energy is defined as the energy use that can be decreased by executing the options that are simple and have a short pay-back period.
In table 1 some examples of these options are listed. These options are all non-investment options.

Other options need small investments e.g.:
- optimizing equipment for heating installations.
- automatical on/off switch for outdoor lighting.
- set point of water temperature for heating governed by outside temperature.
- high efficiency fluorescent lighting.
- re-use of waste water streams for cleaning purposes.
- covering of warm water basins when not in use.
- peak shaving equipment.
- re-use of wood and paper waste.

The options mentioned are all management options. Good management will motivate the employees to use the existing equipment in the most efficient way.
By screening the audit results on these options we can extrapolate the potential of the energy-waste market.
Besides the grant scheme, more information is available from 20 sectoral audits and from the busproject SVEN uses for one-day on site audits.
The result of the screening and the extrapolation to the energy use of today is listed in table 2.
In industry the potential energy conservation is split-up between the potential in savings in comfort and the potential in process.
All these savings can be reached by simple means. They are

the 'good-house keeping' option.
The data provided to us by the consultants was insufficient
for the process part, so here we rely on our own experience
with sectoral audits and bus audits. Based upon this 5% of
the energy used in process can be conserved by implementation of the options listed in table 1. These savings are
all related to final energy use.
To convert the savings to CO_2 reduction emission factors
from the energy study centre are used.
To convert the energy use in PJ into Ktons of CO_2 emission,
the energy mix per sector has to be known.
The energy mix is in this paper is regarded to be supplied
from sources: electricity and natural gas.
One needs to know what the part of electricity is in the
final energy use figure. For the industrial energy use this
figure could be obtained from the Dutch Statistical Bureau,
for the public sector we used our own data. The result for
the Dutch situation are presented in table 4.

The audits from which the figures are presented, are
executed by technical oriented consultants. This implies
that the options regarding management and human behaviour
receive less emphasis than they should. Still it is essential that the managerial options are executed first for a
good organization and motivated employees are of major importance for succesful implementation of technical options.

From discussions with fellow organizations in the
E.E.C. in general the same situation emerged.
In the United States and Canada we expected the waste mar-
ket to be bigger than in Europe, due to the relative low
energy prices and a longer tradition of mechnanisation.
This was confirmed by discussions with their organizations.

In less developed countries the awareness for energy conservation may be greater, but the low efficiency of the
equipment may counteract the possible results of this.
The CO_2 reduction potential in 'good housekeeping' is
around 15% in The Netherlands; the indication is that this
is a global figure.
The implementation of the options for energy conservation
that need investment will yield about the same amount of
CO_2 reduction. But this implementation is dependent upon
the energy price level.

CONCLUSION

The CO_2 emissions can be reduced by 1/8 with simple
energy conservation options. The options are economically

attractive even with the current low energy prices.
Energy conservation is mostly connected with technical measures. If the managerial and human behaviour options are
not implemented before or at the same time, the maximum
possible effect will never be reached. The first step to
reduce the CO_2 emission is to promote energy conservation
by simple means.

TABLE 1 : Energy conservation options non-investment/low investment
- lowering of ventilation in existing buildings
- lowering of temperature during heating season
- installment of thermostatic radiator valves
- lowering of temperature set point during off working hours
- pilot flame off during long absence
- limiting of air-moisture control in buildings
- lowering of air velocities so that temperature can be lowered
- constructing buildings with a high isolation index
- ventilation controlled by needs
- in combined boiler installations: let one boiler run at maximum capacity by shutting off other boilers
- install more controlled heater groups
- installment of room temperature control in the right place
- wind compensation on boiler temperature set point
- warm water supply shut off during non-working hours
- re-adjustment of radiators after insulating the building
- outside temperature used for set point boiler water temperature
- isolation of boilers and boiler equipment (pipes, valves etc.)
- installment of H.E. boiler
- isolation of heat transport pipes
- pilot flame replaced by electric ignition
- optimalisation of fuel/air mixture
- defrosting of refrigerator equipment
- closing curtains to diminish heat loss
- low temperature washing programs
- replacing electric resistance heating by fossil fuelfired stoves
- cooking on gas in stead of electricity
- stop boiler pump outside the heating season
- use of H.E. light appliances
- replacement of lamps by fluorescent lighting
- cooling of buildings during night with outside air
- installment of outside sun shades (trees)
- use of sun boilers for room heating
- critical adjustment of lighting level
- H.E. general lighting
- use of CO_2 from boiler in green houses
- use of equipment heat losses for room heating.

TABLE 2

Sector Code	Name	Energy use (PJ)	Potential savings (PJ)	Needed investments f x 10^6
20/21	Food processing	78.6	9.05	317
22/24	Textiles	7.0	0.344	5.9
26	Paper industry	29.7	1.21	30.3
29/31	chemical industry	213.2	17.872	960
32	construction materials	37.7	1.672	72.5
33/35	metal industry	155	6.358	220
36/39	other industry	15.1	0.654	53
50/52	civil construction	10.6	0.954	31.8
61/62	wholesale trade	18.9	1.769	33.7
65/66	retail trade	26.4	1.462	65.8
67	hotel and catering industry	20	1.94	56.8
68	repair industry	5.9	1.5	93.8
70/76	transport- buildings	10.8	0.624	p.m.
	automotive fuels	325.9	32.59	p.m.

Sector Code	Name	Energy use (PJ)	Potential savings (PJ)	Needed investments f x 10^6
81	banks	5.5	0.601	38.9
82	insurance comp.	1.4	0.142	6.9
83	real estate	1.3	0.26	15.6
84	services	10.5	0.9	64.5
90	government and defence	60	9.4	351
91	churches	4	0.52	45.7
92	schools	21.1	3.04	192.6
93	hospitals	21.7	1.28	70.7
94	social services	15.3	1.12	1.46
95	social cultural in-stitutions	3.4	0.37	30.26
96	sports and recreation	15	1.56	113.5
97	research in-stitutes	5	0.37	23.6
98	other common services	5.4	0.11	3
00	residential sector	468.7	93.75	p.m.

TABLE 3   ENERGY PACKAGE IN DIFFERENT SECTORS

Sector Code	Name	% Electricity	Average GJ price in Dfl	CO_2 Emission (kg/GJ)
20/21	Food processing	17.3	12,-	79.8
22/24	Textiles	20	12,70	83.8
26	Paper industry	27	12,70	83.8
29/31	Chemical industry	14	11,-	75.5
32	Construction materials	20	12,70	83.8
33/35	Metal industry	25	14,-	90.7
36/39	Other industry	27	12,70	83.8
50/52	Civil construction	27	12,70	83.8
61/62	Whole sale trade	32.3	20,-	100
65/66	Retail trade	38.7	22,50	109.6
67	Hotel and catering industry	27.7	18,50	94.4
68	Repair industry	27.7	18,50	94.4

Sector Code	Name	% Electricity	Average GJ price in Dfl	CO_2 Emission (kg/GJ)
70/76	Transport buildings remises automotive fuels	39.9	22,85	111.3
81	Banks	31.6	20,-	99.8
82	Insurance comp.	38.3	22,-	109
83	Real estate	19.4	15,50	83
84	Services	30	19,30	97.6
90	Government and defence	43.9	26,70	116.8
91	Churches	16.9	15,50	79.5
92	Schools	16.4	15,30	78.8
93	Hospitals	32.3	21,90	100.8
94	Social services	24.9	18,80	90.5
95	Social/Cultural institutions	31.7	21,65	100
96	Sports and recreation	29	20,50	96.2
97	Research institution	44.4	26,90	117.5
98	Other common services	29.4	21,55	96.8
00	Residential sector	20	21,-	83.8

TABLE 4A. Potential savings of CO_2 emissions by simple means in The Netherlands.

Sector	Energy use (PJ) 1987	Potential Energy savings (PJ)	CO_2 Emission (K ton)	Reduction CO_2 Emissions (K ton)	%
residential sector	468.7	93.75	39,277	7852.5	20
Agricultural sector	138	6	pm	336.6	
Industry	535.1	37,167	43,931	2973.9	6.7
Transport sector (incl. private cars)	336.7	33,214	24,717	2421.2	9.7
Trade	81.8	7,625	8,113	741.7	9.1
Private services	18.7	1,903	1,835	184.9	10
Public sector	150.9	17.77	15,454	1850.5	11.9
Totals	1729.9	197,749	133,327	16361,3	12,3

The mentioned sectors cover about 80% of the final energy use. Excluded are power generation, water supply and polder drainage.

TABLE 4B REDUCTION OF CO_2 EMISSIONS BY SIMPLE MEANS IN DIFFERENT SECTORS.

Sector Code	Name	CO_2 emission (K ton)	Reduction K ton	%
20/21	Food processing	6,270	722,2	11.5
22/24	Textiles	586	28.8	4.9
26	Paper industry	2,490	101,4	4
29/31	Chemical industry	16,100	1349,9	8.4
32	Construction materials	3,160	140,1	4.4
33/35	Metal industry	14,060	576,7	4
36/39	Other industry	1,265	54.8	4.3
50/52	Civil construction	888	79.9	8.9
61/62	Wholesale trade	1,890	176,9	9.4
65/66	Retail trade	2,890	160,2	5.5
67	Hotel and catering ind.	1,888	183,1	9.6
68	Repair industry	557	141,6	25.4
70/76	Transport, buildings	1,200	69.5	5.8
	other	23,517	2351,7	10
81	Banks	549	60	10.9
82	Insurance comp.	153	15.5	10
83	Real estate	108	21.6	20
84	Services	1,025	87.8	8.6
90	Government and defence	7,008	1,098	15.7

Sector Code	Name	CO_2 emission (K ton)	Reduction K ton	%
91	Churches	318	41.3	13
92	Schools	1,663	239.6	14.4
93	Hospitals	2,187	129	5.8
94	Social services	1,385	101.4	7.3
95	Social/Cultural inst.	340	37	10.8
96	Sports and recreation	1,443	150	10.4
97	Research institution	587	43.5	7.4
98	Other common services	523	10.7	2
00	Residential sector	39,277	7852.5	20
		133,327	16024.7	12

Totaal CO_2 emission in the Netherlands aprox 150000 kton.

ENERGY CONSERVATION FOR A LONG TERM, SUSTAINABLE ENERGY POLICY

H.Y. BECHT and J.P. VAN SOEST
Centre for energy conservation and environmental technology
Oude Delft 180
2611 HH Delft
The Netherlands

ABSTRACT. This article focuses on the perspectives of energy conservation in a 'low CO_2' energy future. It shortly reviews the possibilities that exist in the various end-use sectors and the costs of realising them. An indication is given of a cost-level below which energy-conserving measures can be considered as cost-effective. This level corresponds to the estimated long-term costs of a clean (zero CO_2) energy supply. Finally an overview is given of the main policy instruments that are to be used to realise these opportunities.

1. ENERGY CONSERVATION, ENERGY EFFICIENCY AND ENERGY INTENSITY

Energy conservation or improvement of energy intensity covers many very different activities and processes, that can be divided into two main categories: improvement of energy efficiency of our activities, and a structural change towards less energy intensive activities.

Energy intensity is defined as the total energy demand divided by the gross national product (GNP). Before the energy crisis in 1973, the energy intensity remained remarkably constant during several decades. Since 1973, energy intensity started to decrease, even more sharply after the second oil price shock in 1979 (See e.g. [1], [2] and [3]). Since 1985, the energy intensity is increasing slightly, especially in The Netherlands. Figure 1 illustrates the development of energy intensity with time.

The energy intensity of the economy is dependent on two factors: The energy efficiency of the activities and the structure of the economy, especially the share of ener-

41

gy intensive activities (products like petrol and cars, and services like long distant tourism) in our consumption and production pattern. Before 1973, these factors worked in opposite directions: energy efficiency of products and production processes generally improved, but the growing share of energy intensive activities in the economy (transport, use of materials) were the reason this improvement did not show up in the figure of energy intensity of the economy as a whole. After 1985 this is again the case. Only in the relatively short period of high energy prices both factors worked in the same direction, leading to a significant decrease in energy intensity.

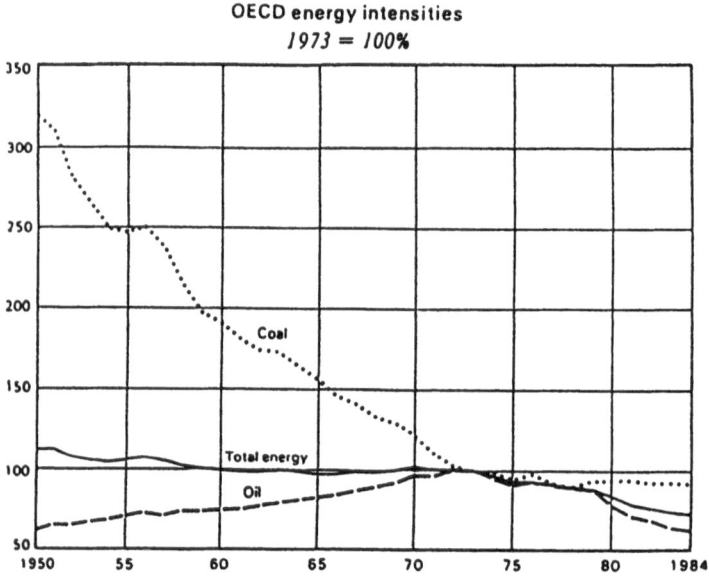

FIGURE 1. Development of energy intensity of OECD-countries

In many publications, energy conservation is limited to improvement of energy efficiency. It will be clear, however, that for a sustainable economic development, improvement of the energy intensity is the crucial factor. In this article, the wide definition of energy conservation as mentioned above will be used: the improvement of energy intensity of consumption by improvement of energy efficiency and by structural change. (Note: on a global scale energy-intensity of consumption equals that of production. On a national or local scale however, the energy intensity of production may deviate considerably from that of consumption. The latter is obviously the more fundamental indicator.)

Energy conservation is only one of the possibilities

to reduce carbon dioxide emissions from the energy supply. Other possibilities are: the use of nuclear energy, the removal of carbon from fossil fuels and storage of the carbondioxide in old oil and gas fields, and the use of renewable sources of energy. Only the last one can be considered as compatible with sustainable development on the long term. The others can play a role during a long transition period.

2. ENERGY CONSERVATION IN THE MAIN FINAL DEMAND SECTORS

2.1. Problems facing energy conservation

In most end-use sectors, a considerable potential for energy conservation still exists. Even at the present low energy prices pay-back periods of only a couple of years are no exception. The penetration of energy-efficient technology, however, is not only dependent on the existence of technical and economic possibilities but many other factors play an important role, such as:
- Lack of knowledge on energy efficient technologies.
- Lack of attention: energy usually forms only a small part of production costs; thinking about reduction requires attention, time and money: energy conservation requires a rather high 'activation energy' itself.
- Requirement of high rates of return on investments in energy conservation. This places energy conservation in an unfavourable position in comparison with energy production, where utilities are accomodated to long payback periods and low rates of return on capital investments.
- The revenues do not always directly benefit the investor; this goes for example for the owners of rental homes and buildings.

In order to have an impression of the amounts of energy involved, the following table (on next page) shows the energy balance (supply and demand) of The Netherlands in 1985.

TABLE 1. Energy balance of The Netherlands, 1985
(source: [4]).
Units: PJ (1 PJ = 31.6x10^6 m^3 natural gas or 24 m.t.o.e)

sector:	coal	oil	gas	electr.	nuclear & other	heat	Tot
final demand							
industry (excl. CHP)	71	34	221	89		110	524
non-energy (feedst.)	10	223	90	11			334
transport		322		4			326
domestic, services & other	10	122	636	117		8	893
total final demand	90	701	947	231		118	2077
transformations							
conventional electricity	150	7	301	-183	51	-11	315
Cogeneration (CHP)	12	26	115	- 19		-87	37
Refineries & others	12	112	26	- 0		-11	139
total primary demand	264	846	1,389	19	51	9	2,568

2.2. Industry

In industry, energy is used as a feedstock (334 PJ) and for processes (heating, pumping, drying, cooling (435 PJ heat and 89 PJ electricity). Relatively small amounts are used for space heating and lighting. Fuel input as a feedstock can be reduced by a decreasing use of materials per unit of end product and by recycling of materials. By recycling, savings up to more than 50% are possible. The reduction of the use of materials per unit of end product expresses itself in a decreasing share of the base-metal and base chemicals in total industry.

Energy for processes (heat and electricity) may also be reduced significantly. Typical measures are [5]:

Measure	saving potential (indication)
- small, low-cost measures	20% (heat and electricity)
- heat pumps	10% (heat, extra use of electr.)
- heat recovery (heat exchangers)	10% (heat, extra use of electr.)
- mechanical vapour recompression	25% (heat, extra use of electr.)

- reversed osmosis   25% (heat, extra use of electr.)
- changing of processes   P.M
  (e.g., application of electrochemistry, catalysts or bio-
  technology)
- more efficient cooling   20% (electricity)
- efficient motors, speed   20% (electricity)
  control
Total: heat: 40% (long term 60%), electricity: 20% (long term 30%)

It may be concluded that by recycling and by more efficient equipment, the energy efficiency of industry can
improve with at least 2% per year. This efficiency improvement will only be obtained as a result of a strong policy
(see paragraph 4). Costs of energy saved range from zero up
to approximately Dfl 12/GJ for heat and Dfl 25/GJ for electricity. Additional to this efficiency improvement, structural changes can be expected from a strong policy. Such a
policy should however be co-ordinated on at least a European scale.

2.3. Transport

Efficiency, occupancy and volume
     Traffic and transport are a fast growing sector of the
Dutch economy and therefore contribute increasingly to the
mondial CO_2 burden. Energy use in 1985 amounted to 326 PJ
(other sources [6], [7] estimate 356 PJ), accounting for
26×10^9 kg CO_2, or 15% of total emissions.
     In traffic and transport there are three main factors
which determine the expected CO_2 emissions for the future:
vehicle-technical efficiency improvements, occupancy ratio
of the means of transport, and the volume development (passenger and ton kilometres). The estimates of the extent of
these factors diverge; furthermore a part of the developments is dependent on the pursued policy. These points will
shortly pass in revue below.

Trend at unchanged policy
     Of the vehicle-technical potential only a part is ex-
pected to be realized autonomously. To an important extent
the low fuel prices and the absence of a policy specifically based on energy efficiency are to blame. Optimistic estimates of the maximum efficiency improvements for cars
2010 amount to approximately 35% [8], [9]. Own estimates
indicate that improvements of some 40% (maximum value) of
the vehicle efficiency of passenger cars are technically
feasible, but only under the influence of a vigorous policy, which is presently lacking. In present circumstances,
20% is a more likely figure.
     Comparable notes can be made with regard to the

freight transport by road: the technical potential of ca. 35% efficiency improvement ([16], [9]) is expected to be unrealistic; even under the influence of a vigorous policy a maximum improvement of 20% seems to be feasible. Without additional measures 10% seems to be a more realistic estimate. Also as to the possible increase of the occupancy ratio and volume developments several estimates have been made (see table 2), depending on what measures will be taken.

Policy-options: technology and occupancy ratio

Proceeding vehicle-technical improvements as well as an increase of the occupancy ratio can be realized by additional policy-measures. An increase of the energy efficiency can be advanced by standards (as is done in the USA) of particularly cars, and/or by an increase of fuel prices. This can result in a relative increase in the number of lighter and smaller cars, a development which will not take place autonomously in a growing economy. In public transport recuperation of brake-energy can be applied.

The occupancy ratio of cars can be increased by 10% [7] by among others an increase of fuel prices, introduction of road-pricing, reservation of traffic-lanes for car-poolers (so-called 'diamond lanes'). Also the occupancy ratio of public transport can be somewhat increased (ca. 10%) by a number of specific measures. For freight transport an additional improvement of the occupancy ratio of 5% is considered to be possible.

The effect of these improvements of the vehicle-efficiency and the occupation ratio on the CO_2 emissions in 2010, as calculated in table 2, is still an increase in energy use and in total emission of CO_2: 18%, instead of 53% growth (trend).

Policy-options: modal split and volume policy

In the long term proceeding vehicle-technical adaptions are thinkable: change-over to methanol and hydrogen as fuel, a relative increase of electrical traction. In the long term, low CO_2 options are available. For the short term a shift of modal split and decrease of mobility are the most obvious paths for a decrease of energy use and thus of CO_2 emissions.

A shift of freight transport from 'road' to 'water and rail' can effectuate a reduction of energy use by ca. 85% per ton-kilometre. For passenger traffic a change from car to public transport means a reduction in energy use of about 35% per passenger kilometre; a shift from car to slow traffic makes the emission of carbon dioxide drop to zero.

As a result of price measures, e.g. a price increase of transport fuels of Dfl 2/liter (current prices being some Dfl 1.5/liter), volume trends would change. Volume effects of such a measure are calculated in table 2.

Conclusions
        Table 2 describes the expected CO_2 emissions of the traffic sector for 2010 at unchanged policy, and as a result of two policy options: enhanced technological developments and improved occupancy ratio (A), and A in combination with volume policy and change in modal split (B).

TABLE 2. CO_2 emissions of traffic and transport in 2010 (as compared to 1985).
ve = vehicle efficiency improvements (%)
or = occupancy ratio improvements (%)
vol = volume development (%)
CO_2 x10^9 kg
Policy options:
(A) = enhanced technological developments and improved occupancy ratio
(B) = (A) + fuel taxes (Dfl 2/liter) + modal split change

	trend				policy A				policy B			
	ve	or	vol	CO_2	ve	or	vol	CO_2	ve	or	vol	CO_2
Pass. car	20	-	48	17.0	40	10	48	11.5	40	10	-10	7.0
Public tr.	-	15	100	1.5	20	25	100	1.1	20	25	200	1.6
Freight tr.												
-road	10	10	80	7.5	20	15	80	6.3	20	15	20	4.2
-rail	-	10	25	0.1	10	15	25	0.1	10	15	125	0.2
-water	10	10	45	1.1	30	15	45	0.8	30	15	65	0.9
Aviation	30	-	240	8.9	35	5	240	7.9	35	5	140	5.5
Other	-	-	50	0.6	-	-	50	0.6	-	-	50	0.6
Total				36.7				28.3				20.0
(1985 = 25.5)												

2.4. Domestic, services and agriculture

Domestic energy
        The amount of energy consumed for domestic use (end user) in 1985 totalled about 500 PJ. The following table shows the distribution of energy-carriers (or forms) over a number of end use categories:

TABLE 3. Energy use in the domestic sector in 1985 (source [10])

source of energy	type of end use					
	space heating	hot water	cooking	appliances	light	total
natural gas	360	53	11			424
oil, LPG and coal	17					17
electricity	1	8	1	31	17	58
hot-water (district heating)	4					4
total	382	61	12	31	17	503

Most important are natural gas used for heating, domestic hot water and cooking, and electricity used by domestic appliances and lighting. Subsequent CO_2 emissions total about 25×10^9 kg excluding electricity (18% of total emissions in The Netherlands) or 35×10^9 kg (25%) including electricity.

Space heating
In [11] trends have been analyzed, concerning heating in existing buildings and new buildings as well as their effects on emissions of SO2, NOx and CO_2. The possibilities to save energy in order to reduce emissions and their costs and gains are also analyzed.

Present policy. With an unchanged policy the now existing number of dwellings of ca. 5.5 milion will grow to approx. 7.2 milion in 2010 (30% increase). Large uninsulated dwellings will disappear and be replaced by smaller well-insulated ones.

Furthermore some 70% of the houses in the renovation programme will be fitted with some form of insulation. The average consumption of gas per house for heating will decrease from ca. 2,000 (in 1985) to about 1,640 m3 natural gas per year in 2010. This effect however, will be totally obliterated by the increase in the number of dwellings. The net result is an increase in the use of natural gas from 11×10^9 m^3 per year to some 12×10^9 m^3.

Existing dwellings. Evaluation of the Nationaal Isolatie Programma (NIP, National Insulation Programme) [12], shows a large part of the existing dwellings to have some form of insulation, but only a small portion has been insulated completely. Only 25% of the technically possible measures have been implemented.

Especially dwellings built before 1974 show bad figures: only 20% have been completely insulated. For dwellings built after 1974 this figure is 47%. For both types of dwellings mentioned there is still a huge potential, which should be tapped in order to further reduce CO_2 emissions. According to [11] the following measures are considered feasible:
- at renovation 90% of the dwelling should be fitted with insulation instead of only 70%;
- the quality of insulation should be increased; the consumption of energy remaining after renovation (including insulation), now estimated to be 80% of the original consumption, could decrease to about 50%. Extra investments are about Dfl 2,500.- per dwelling (excl. VAT).
- Some 100,000 dwellings per year should be retro fitted with insulation: the use of energy could then drop to about 70% of the original figure. Costs are about

Dfl 1,500.- per dwelling.

These measures could realize a reduction of the average annual consumption of energy per dwelling to about
1,400 m^3. Application of high efficiency heating systems
and heat recovery may further lower this amount by 200 m^3
per dwelling. Cost of energy saved is about Dfl 10/GJ.

New dwellings. Substantial proceeding energy saving at
acceptable cost effectiveness is also possible for new
dwellings. In practice it has been proved that a decrease
of the average natural gas consumption for space heating
per dwelling up to ca. 600 m^3 is feasible (extra investment
ca. Dfl 3,500.- per dwelling). Cost of energy saved is also
about Dfl 10/GJ, because of the longer lifetime of the investments.

Existing and new dwellings. If the policy for the new
dwellings (a decrease to 600 m^3) would be realized, in combination with the above described energy saving in existing
dwellings, the total natural gas consumption could drop to
about 7.8x10^9 m^3 in 2010. The 'energy intensity' of the
dwelling stock would decrease to 55% of the present stock:
1,100 m^3 natural gas/dwelling in 2010 versus 2,000 m^3 natural gas/dwelling in 1985.

Domestic hot water and cooking
The use of energy for domestic hot water and cooking
is only a relatively small portion of the gas used in the
dwelling. According to estimates another 20% could be saved
through the use of appliances and heaters without pilot
flames and by application of solar hot water systems. The
saving potential of solar hot water systems is estimated
to be about 12 PJ in existing dwellings and another 12 PJ
in new dwellings [13]. At present the average pay-back period is longer than 10 years.

Electricity
The domestic sector is one of the few sectors of which
the use of electricity, spread over its applications, is
well-documented. Such information is essential for estimating the saving potential. The use of electricity for domestic application (in 1985 ca. 58 PJ in The Netherlands)
increases yearly with several per cent, most probably under
the influence of low electricity prices and increased in-
comes, and in spite of a yearly average of a good 1% efficiency increase of the appliances, thanks to the ongoing
technological improvements. Important consumption items are
lighting, cooling and freezing, washing and heating and
domestic hot water preparation (including central heating
pump).

TNO [14] estimates the autonomous improvement of efficiency up to 2000 (without supplementing policy) at 12% for households. Own estimates [15] come down to a technical economic potential of 40% increasing to 60% on the long term.

Demographical and economic developments (increasing number of households, an increasing penetration of electrical appliances per household) seem to nullify the improvements of energy efficiency (certainly the autonomously acting part of it) so that in the future an increasing domestic use of electricity should be counted on.

Total domestic energy consumption

Through a vigorous promotion of savings on electricity, reduction of the domestic use of electricity with 10-20% is possible, particularly lighting (savings of about 70% are possible), laundering at low temperatures and improvement of efficiency of other domestic appliances. The domestic use of energy will then be some 323 PJ for heating (mainly natural gas), cooking and hot water, and for electricity 46 PJ.

Service sector

Total energy use in the service sector in 1985 was 140 PJ (heat) and 45 PJ (electricity). New office buildings have been realised that consume only a third of the heat and electricity that is used by conventional buildings. Retrofitting of old buildings can save approximately 30% on energy use. Insulation, efficient lighting and air-conditioning systems are the most important measures. Cost of energy saved may be estimated at Dfl 5-8/GJ (Dfl 0.15-0.25/m³ gas equivalent) for heat (gas) and Dfl 10-25/GJ (Dfl 0.04-0.10/kWh) for electricity.

Relatively low energy prices will prevent realisation of these opportunities and rapid growth of the service sector will lead to a moderate increase of energy use. With strong incentives for energy conservation an improvement of energy efficiency of 40% by 2010 seems feasible. Assumption of a net growth rate of building volume of 1% per year leads to a possible reduction of energy use with 1% per year, both for heat and electricity.

Agriculture

Energy use in agriculture is mainly concentrated in horticulture under glass (greenhouses). Total energy consumption in 1985 was 83 PJ (heat) and 10 PJ (electricity). A significant potential (more than 50%, [5]) for energy conservation still exists, especially by crop intensification, higher yields (value) per square metre, and by lower heat input per square metre by using heat shields and better glazings. The use of electricity is growing fast because of the application of greenhouse lighting, lengthening the growing season. With strong incentives for energy conservation this growth may be slowed down but not stopped.

Electricity generation

Various estimates, among others [16] indicate a significant untapped potential for cogeneration, the combined generation of heat and power (CHP), up to 4,000 MW additional to the existing 2,000 MW (industry: 2,000 MW, agriculture: 1,200 MW, domestic and services: 800 MW). Realisation of this potential leads to an energy conservation of 70 PJ per year at a cost of approximately Dfl 7/GJ.
Figure 2 shows the CO_2 emissions of various electricity generating technologies. Realisation of the potential would implicate that half of the present electricity production is generated in cogeneration plants.

EMISSION FACTORS OF CO_2 RELATED TO ELECTRICITY PRODUCTION
(EXCL. EMISSIONS RELATED TO CAPITAL INVESTMENT)

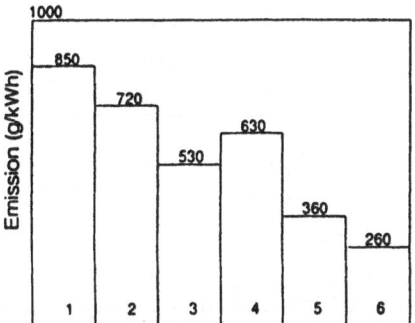

Mode of production
1 Coal conventional (e = 40%)
2 Coal, IGCC (e = 47%)
3 Coal, CHP (e = 64%)
4 Oil conventional (e = 42%)
5 Gas, CC (e = 55%)
6 Gas, CHP (e = 75%)
(e is effeciency)

FIGURE 2. CO_2 emissions of electricity generation technologies

2.5. Summary of the potential for energy conservation

The improvement of energy efficiency in the various end use sectors can be approximately 2% per year during the coming 20 years. Structural changes are possible in the transport sector and in industry. These may add 1% per year to the improvement of energy intensity of our economy, which is thus maximally 3% per year. For comparison: during 1973-1985 the reduction of energy intensity was only 2% per year.

TABLE 4. Minimum estimate of energy demand in The Netherlands in 2010, as a result of a 'sustainable energy policy' in a 'coal, average price development scenario', [5],
[17].

sector:	volume growth	efficiency gain		forms of energy		
		heat	electr.	fuels	electr.	total
	%/year	%	%	PJ	PJ	PJ
final demand						
domestic	1,4	45	40	323	46	369
services	1,5	40	40	113	36	149
agriculture	2	50	0	66	15	81
industry	2,5	40	25	428	122	550
feedstock	1	50	(recycling)	197	23	220
transport	1.2	35	-	246	10	256
others	-	-	-	108	4	112
total final demand				1,481	256	1,737
transformations						
conv. electricity				327	-147	180
Cogeneration				156	-109	47
refineries & others				110		110
total primary demand (2010)				2,074	0	2,074
(compare 1985:				2,568		2,568)
('official' projection 2010 [17]				3,400		3,400)

The improvement of energy intensity by 3%/year is in
strong contrast to the present stabilisation or even increase of energy intensity. It may be concluded that an
extremely strong and effective conservation policy is required to obtain this reduction, which is technically feasible and cost effective (cost of energy saved lower than
Dfl 12/GJ for heat and Dfl 25/GJ for electricity).

In order to illustrate the dominant influence of the
growth of production on energy demand, table 4 shows the
energy balance of The Netherlands 2010, as the combined
result of the 'official' projections of economic development [17] and energy conservation as described in this
chapter: even with maximal energy conservation only a
slight reduction of total energy demand can be expected.
Possibly this conclusion is too pessimistic: a strong environmental policy will in our opinion lead to a lower
growth of production in the traditional sense, because the
structural changes towards a low input economy reduce the
share of the traditional (and present) growth generating
sectors in the National Income.

3. COSTS OF AN ENERGY SUPPLY WITHOUT CO_2

Cost estimates of energy supply options without CO_2 emissions are highly speculative. These estimates, however, are more reliable than those of damage to be attributed to the use of fossil fuels. Presumably the cost of a 'clean' energy supply will also be lower than the damage to be expected from the greenhouse effect.

For energy policy an indication of the future costs of a 'clean' energy supply is very important: it is for example not very useful to stimulate forms of energy conservation that are more expensive than the costs of clean energy. In [18] the following overview is given of several options to remove CO_2. These are also discussed in other contributions to the conference 'Energy and Climate'.

TABLE 5. Summary of potential and minimum estimates of the long term costs of energy supply options without CO_2 emissions (source: [18])

option:	potential costs PJ/yr.	Dfl/GJ
1. inland production of energy from renewable sources (photovoltaic + wind, incl. continental shelf)	ca. 900	17 el
2. production of hydrogen from surpluses of 1.	-	25 H_2
3. production of solar heat and biogas	ca. 200	25 th
4. production of electricity from breeder reactors	large	>23 el
5. production of hydrogen from surpluses of 4.	large	>25 H_2
6. import of liquid fuels from biomass	large	>15 th
7. import or hydrogen from solar energy in (sub-)tropical areas	large	>15 H_2
8. hydrogen produced after removal of CO_2 from fossil fuels and storage of CO_2 in old oil/gasfields (minimum storage capacity ca. 50 yrs.) (H_2 from natural gas at present production cost of natural gas costs approx. Dfl 7/GJ H_2)	2500	15 H_2

el = electricity
H_2 = hydrogen production
th = thermal

It may be concluded, that the cost of a 'clean' energy supply will be at least Dfl 15/GJ. This corresponds approximately to the oil price level in the beginning of the 1980s. A rational energy policy will try to realise all options for energy conservation which are cheaper than this level, which corresponds to 'the Long Term Avoidable Costs' or LTAC. Generally the options described in chapter 2 fulfil this condition.

4. POLICY INSTRUMENTS FOR A SUSTAINABLE ENERGY FUTURE

4.1. Introduction

In most economic sectors contemplated a similar development can be detected: there is an extensive technical saving potential, as large as tens of per cents; only a modest part is realized autonomously. An extensive supplementary policy instrument is necessary to 'grasp all' concerning the improvement of the energy intensity of the economy.

Nevertheless, even by implementation of vigorous policy instruments the effects of economic and demographical trends on the total energy consumption and consequently the CO_2 emission are of decisive importance. In the decades to come 'sustainable development' as advocated by the Brundtland committee is out of the question. An important factor that blocks the sustainable development is the fact that at present fossil fuel prices do not include environmental costs and future risks; also long term scarcity is not expressed in the present market prices for fossil energy. These observations compel the mapping out of a 'sustainable energy policy' (see also [19]), which will be further worked out in the following paragraphs. Main elements are price and tariff policy, standards, the task setting of the utility companies, research and development, as well as transfer of knowledge and information.

4.2. Price, tariff and subsidy policy

The analyses of the energy intensities of the economies of the Western Countries have clearly proved a strong influence of the energy prices [1]. Regarding the issue of the promotion of a sustainable development, an economy with a low energy intensity, the first item to be thought of concerning the effectiveness must be price and tariff policy. It is clear, that such measures should preferably be taken in a European context, even though a somewhat limited internal price policy in a number of areas is positively possible.

The basic principle for tariff setting that can promote the sustainable development, is a price policy that originates from long term avoidable costs. This means that the users of utility services, the use of which, from a social point of view, must be decreased as much as possible, (e.g. energy supply, waste disposal, water treatment) in the present situation are already presented a tariff which corresponds with the minimum of costs to be expected in the long term. Such a price setting strongly stimulates reduction of the use of this utility service. For the energy supply these long term avoidable costs (see chapter 3) are computed at the basis of the starting point that in the future a clean, CO_2 low energy supply is possible and necessary [19] and estimated at Dfl 15/GJ at least. As a first step in that direction one can think of the increase of the (modest) investment levy on natural gas for small users, with ca. Dfl 0.02 per m^3, a levy of e.g., Dfl 0.005 per kWh on electricity for small users, and the increase of fuel taxes for transport (with at the same time a decrease of fixed costs: variabilisaton). With the proceeds of these levies investments in energy conservation and sustainable energy can be stimulated. With proceeding increases effects on the income distribution deserve proper attention and perhaps a compensating policy must be considered, preferably in an 'energetic form', such as abolishing the standing charge for low income groups, or the establishment of special energy programmes. Another option is an overall decrease of the (low) VAT tariff.

An increase to the previous mentioned level of Dfl 15/GJ should lead to an extra tax income of Dfl $10-20x10^9$ at the present world market prices. The low amount is obtained through application to small users and small single use. The high amount is obtained through application to all forms of fuel consumption. In this framework a levy in proportion to CO_2 emission is relevant, as has already been pleaded in a number of studies.

The World Watch Institute [20] pleads a levy in the magnitude of Dfl 2.25/GJ for coal and Dfl 1.75/GJ for oil. The UPI [21] proceeds substantially by its proposals for an 'Eco tax' on fossil fuels in the magnitude of several guilders per GJ, yielding an energy price level close to the aforementioned long term avoided costs of some Dfl 15/GJ.

The levies could be decreased at a increasing world oil price. Levies to such extent are, certainly for large scale industrial use and international transport, only feasible in international (at the least EC) relations. Next to levies other stimulating financial or restraining instruments are possible. Particularly stimulating tariffs should be mentioned for clean and economical electricity (co generation, renewable energy) delivered to the grid, and such a natural gas tariff for cogeneration that this form of energy saving can compete with central production. With respect to traffic the abolishment of fiscal measures that increase mobility is a good option, as seen next to variability and the increase of fuel tax. Particularly the abolishment of the deemed travel expenses allowance, the introduction of taxing of car allowances rising above the variable car costs, and heavier taxing of private use of company cars.

Subsidies promoting the use of clean and efficient energy technology are the counterpart of price stimuli through energy carriers or energy forms that aim to make investments in savings inherently more profitable. Many of the technical measures described in previous paragraphs could be promoted this way. These subsidies could be paid for by the above mentioned taxes. The feasibility of subsidy regulations is especially substantial when applied to relatively simple, easy to describe technologies such as cogeneration plants, solar cells and windturbines, efficient lighting, and pump controls.

4.3. Energy efficiency standards

Standards and regulation for energy use are especially suitable instruments for those applications where regulations do not seriously hinder the freedom of choice of the consumer (or industry). Possibilities are to be found in dwellings, utility buildings and in domestic appliances and cars. Standards for the maximum use of energy are very well possible and have already been applied in several communities. For dwellings one can think of a standard of about 850 m^3 natural gas or lower, for offices a maximum heat demand of 1,5 m^3 natural gas per m^3 building volume. Furthermore, the use of solar systems for domestic hot water or enlarged solar systems (to contribute to space heating as well) may be dictated.

The spread of energy efficiency of domestic appliances is large, it is possible to set standards, based on the most efficient 10% of the appliances. This does not necessarily have to be an obligatory standard, enhancing energy conscious buying is also possible by giving a hallmark or label to the most efficient appliances and to add to this a subsidy (possibly to be paid through the utility company). This subsidy could sometimes be partly paid for by the savings on costs by the utility company (less power capacity needed).

4.4. Task setting of utility companies

Utility companies could play a key role at the introduction of energy saving techniques. Important is a financial and organisational separation between central production of energy on the one hand and the distribution of energy, including local energy efficient generation and energy saving, on the other. The energy distribution utilities must be set a clear task in order to ensure the availability to the consumer of energy services for the lowest possible social costs; so called 'least cost planning'; a concept more and more accepted in the USA. Energy services means: heating, cooking, light and power; in short all the functions a consumer needs. This approach is a serious change as opposed to the traditional view that a utility only supplies kilowatthours and gas and does not 'look behind the meter' of the consumer. With this new philosophy the distribution utilities can expand much beyond their conventional task: information. The services of the utility company stretch from advising on new appliances, via performing energy audits, to financing and leasing of low energy equipment (e.g., solar systems) and investing in wind energy and cogeneration. In order to succeed the tariff between central production facility and the distribution utility should be such that energy saving becomes really attractive to the distribution utility. Furthermore, the distribution utilities should have such tariffs, that energy saving by consumers is stimulated.

4.5. Research, technology development and transfer of knowledge

To suppliers of low energy equipment the perspective of an existing or future market has always been the most important drive towards innovation and renewing of equipment. The creation of favourable conditions for the introduction of low energy technology through ways described in previous paragraphs seems to be a conditio sine qua non for a sustainable energy policy.

As a supplement to this, it is nevertheless desirable to create and to intensify programmes for research and development of technologies that contribute to a sustainable energy supply. Important points of focus are:
- Energy saving: detailed evaluation of the use of energy per sector and per activity, and research and development of promising saving options.
- Renewable energy sources: research and demonstration of renewable energy systems, development of new generation systems, attention to the linking of renewable energy sources to each other, and bio-fuels.
- Other conversion systems: research, development and demonstration of conversion systems based on fossil fuels, which can act as a transition towards a sustainable energy supply, especially coal gasification, natural gas conversion and the possibilities of bio-fuels and hydrogen.
- Hydrogen: probably one of the most important secondary (interchange) fuels in a low CO_2 energy supply.

5. CONCLUSIONS

A cost-effective improvement of energy intensity of the economy with some 3% annually seems to be feasible, in case of a strong policy aiming at energy conservation. Pri-

58

cing of energy at a level of Long Term Avoidable Cost (LTAC), at least 15 Dfl/GJ, will be a cornerstone of such a policy. Seen against a background of economic projections for 2010, this improvement can only lead to a decrease of energy use of some 20% as compared to 1985. The reduction of CO_2 emissions which most likely is necessary (80%), can only be achieved when additional measures are taken. These consist of technical measures for CO_2 removal and further adjustments of economic development towards a low input economy.

REFERENCES

[1]   International Energy Agency. (1987) Energy Conserva-
       tion in IEA-countries. Paris.
[2]   Energy Efficiency; the good and bad news. Energy in
       Europe, 12, March 1989. Article based on research by
       the West-German Frauenhofer Institute: Energy conser-
       vation indicators (1987).
[3]   Various OECD-statistics.
[4]   CBS energy statistics (De Nederlandse energiehuishou-
       ding), The Hague, 1987.
[5]   Blok, K. (1989) Data acquisition of energy conserva-
       tion techniques for the year 2010. Dept. of Science,
       Technology and Society, Univ. of Utrecht, June.
[6]   Bureau Goudappel Coffeng. (1987) Energiebehoefte in de
       transportsector. Fase 1.
[7]   Bureau Goudappel Coffeng. (1987) Energiebehoefte in de
       transportsector. Fase 2a: het jaar 2010.
[8]   Nationaal Milieubeleidsplan, mei 1989. (National en-
       vironmental policy plan, May 1989).
[9]   Tweede Structuurschema Verkeer en Vervoer. Den Haag,
       1988.
[10]   Krekel, N.R.A., P.A.M. Berdowsky en A.J. van Dieren.
       (1987) Duurzame energie, een toekomstverkenning. Rot-
       terdam.
[11]   Oltheten, M. en H. Sips. (1989) Emissiebeleid in de
       woningbouw. CE, Delft, April.
[12]   Berbée, F.P.J. en J.M.J.F. Houben. (1989) Evaluatie 10
       jaar NIP. Gas, no. 2, February.
[13]   Van den Haspel, B. en J. van den Doel. (1989) Het
       emissiemodel ENIMOD; emissiepreventie in relatie tot
       emissiebestrijding. CE, Delft, May.
[14]   TNO (1987) Elektriciteitsgebruik in huishoudens en
       bedrijven. Samenvattende analyse van ontwikkelingen en
       besparingspotentiëlen tot het jaar 2000. Den Haag.
[15]   Dinkelman, G.H. (1988) De markt achter de meter. Ver-
       slag van een onderzoek naar de bevordering van elek-
       triciteitsbesparing door energiedistributiebedrijven.
       Delft.

[16] WKK in perspectief; CE & Univ. of Utrecht, March 1989.

[17] ESC - Nationale energie verkenningen. Petten, Sept. 1987.

[18] Becht, H.Y. (1989) CO_2-emissies van lange-termijn energiescenario's; emissiereductie in de KWW-scenario's. CE, Delft, July.

[19] Becht, H.Y. (1987) Duurzaam energiebeleid. CE, Delft.

[20] Flavin, C. and A.B. Durning (1988) Building on success; the age of energy efficiency. World Watch Institute, paper 82. March.

[21] Teufel, D. et al (1988) Oekosteuern als marktwirtschaftliches Instrument in Umweltschutz. Umwelt- und Prognose Institut, Heidelberg.

PROSPECTS FOR CARBON DIOXIDE EMISSION REDUCTION

F.M.J.A. DIEPSTRATEN, W. VAN GOOL
Energy Science Project, State University Utrecht
Croesestraat 77ᵃ
Utrecht
The Netherlands

ABSTRACT.   The importance of exergy analysis as a tool to investigate the improvement potential of the energy productivity of society is demonstrated. It is shown that the sectors Household & Other and the sector Industry have the best potential for improvement of the energy productivity and for reduction of the carbon dioxide emission.

The major aspect of the carbon dioxide problem concerns the size of the carbon dioxide emission: the amounts involved are roughly a factor 100 larger than those leading to acid rain (sulfur oxide and nitrogen oxides).

Based upon exergy matching of demand and supply, six applications are mentioned which can lead to lower carbon dioxide emissions. No single option has a relevant potential for reducing the carbon dioxide emission on a short term, while maintaining the present production. The only short term option to reduce carbon dioxide emissions is to reduce fossil fuel using productions.

A number of measures is collected in a simple model to investigate whether or not they can lead to stabilization of carbon dioxide emissions by 2015, in a period of growing economic activity. The measures proposed above cannot accomplish such a stabilization. In a situation of growing economic activity more drastic measures will be necessary.

1. INTRODUCTION

In the last decades many studies have been performed to predict the effects of increasing carbon dioxide concentrations in the atmosphere. Although the results of these studies are not yet conclusive with respect to the time scale and the impact of such an increase, it is possible that a reduction of the carbon dioxide emission will become mandatory in the future. Therefore, it is important to be prepared for such a situation.

Many scenarios have been developed to estimate results of technical or economic actions with respect to the carbon dioxide emissions. Several reviews, books, and papers have been published [1-4]. Recently a study of the options in the Dutch society has been reported [5].

The present paper is based upon a continuing effort to analyze the energy losses in society and the options to improve the energy productivity in terms of exergy. The official statistics of nations are generally reported in terms of enthalpy, which pertains just to the amounts of energy involved. The exergy refers to the quality of the energy demand and supply. Exergy analysis is the only methodology to establish where real energy losses occur in society.

The importance of the use of the exergy description in analyzing the carbon dioxide issue originates from the fact that this is the most impelling problem mankind faces, if the problem really exists. The amounts of carbon dioxide emitted are staggering, equal to or even in excess of the amounts of fuels used. This means that the problem cannot be solved by some energy conservation or some (increase of) fuel tax: it will be necessary to increase the exergy efficiency of society fundamentally.

In this preliminary report of recent research the scope of the problem will be illustrated with a few down-to-earth examples. Besides illustrating the exergy concept, an exergy analysis will be given of some aggregated sectors. This analysis shows the direction of where to go in the future. Some technical options are discussed to improve the exergy productivity. It will become obvious that the implementation of these options will take many decades. Furthermore, there is no single option that can turn the carbon dioxide flood. Thus, the discussed options are taken together in a straightforward model to see whether or not their collective application can lead to stabilization of the carbon dioxide production of The Netherlands by 2015.

2. EXERGY AND CARBON DIOXIDE

2.1 Exergy analysis

In the introduction the concept of exergy was mentioned. Although the necessity to use the exergy description for the detailed analysis of energy losses in society is demonstrated in modern energy science literature, the use of the exergy concept has not yet reached the level of policy making. A number of simple examples will clarify the meaning of exergy.

Exergy is the energetic quality of energy and material resources: it is the amount of high quality work which can be obtained from the resource when it is converted reversibly to a stable reference state. For fossil energy carriers there is not much difference between energy (enthalpy) and exergy. The ratio of exergy to lower heating value for most primary fossil fuels lies between 1.03 and 1.06. No significant error is made when the enthalpy is used instead of exergy.

Exergy (availability in American literature) is analogous to the Gibbs free energy as used in equilibrium thermodynamics:

Gibbs free energy $= G = H - T * S$

Exergy $= B = H - T_0 * S$

in which H and S are the enthalpy and entropy, T the temperature and T_0 the reference temperature. The use of Gibbs free energy in a multiphase system requires that all phases have the same temperature and pressure. Exergy can be used in steady-state multiphase processes also when the intrinsic parameters are different. Gibbs free energy uses the elements at standard temperature and pressure as reference. Exergy refers to a socalled environmental reference system in which compounds that are ubiquitous in the environment (air, water) posses zero exergy.

The most striking differences between the use of exergy and enthalpy occur when the appreciation of heat is involved. Enthalpy just gives the amount of heat. Exergy gives the amount of high quality work that can be gained from this heat. Large amounts of enthalpy are lost at the end of a power generation cycle, where the low temperature heat is discarded in cooling-water into the environment. From the exergy point of view, this loss is practically zero. Nothing can be gained from this flow, no work can be done by it. The exergy losses occur on the high temperature side of the process, in spite of the fact that the enthalpy efficiency of that part of the process is quite good.

This difference in appreciation of heat in terms of enthalpy and exergy exists also with respect to the central heating systems of houses. The enthalpy efficiency of the newer heating equipment can be as high as 90%. Exergetically, central heating systems are just about the worst invention ever. A high quality fuel (in The Netherlands mainly natural gas), capable of producing heat of about 2000 °C, is used to generate hot water of about 90 °C, which in turn is used to keep a room at 20 °C. The exergy efficiency of this process is below 10%, since the temperature of the generated heat (room temperature) is only slightly above the ambient temperature. The quality, initially stored in the fuel, is destroyed.

When an increased level of carbon dioxide really has the disastrous consequences that are sometimes predicted, improving the exergy efficiency of society considerably will be imperative. Exergy analysis is the only tool to establish these losses.

When trying to diminish the use or the production of a certain commodity, it is worthwhile to start with the largest contributor. We have divided energy users in several sectors (as used by the Dutch Central Bureau of Statistics). The improvement potential of a sector can be evaluated by considering two factors: the mean exergetic efficiency of a sector and the total amount of exergy used. We define the Process Improvement Potential by:

PI-Potential = (1 - exergy efficiency.).(exergy use)

This index has been used in earlier papers to validate unit operations in processes [6-7]. A high exergy input combined with low exergetic efficiency provides a high improvement potential. To facilitate comparison between sectors, the highest PI-Potential is set on 100. The other sectors are compared to this standard setting leading to the Process Improvement Index.

Correspondingly, a carbon dioxide improvement potential and index (CDI-index) can be defined, using:
CDI-Potential = (1 - exergy efficiency.).(CO_2 -production)

The first factor again represents the possible improvement in process efficiency. The second factor is the sum of CO_2 produced by the various energy carriers according to their specific emission factor. The emission factors used in the calculations are given in table 1. The result of the indexing is given in table 2.

Conclusion: using exergy analysis it is shown that the sectors Household & Other and Industry have the best potential for improvement of the energy productivity.

TABLE 1. Emission factors.
Source: [8,9].

Energy carrier	Emission factor (kton CO_2 /PJ)
coal	94
oil	75
natural gas	56
gasoline	73
diesel	73

TABLE 2. PI- and CDI-index for energy users in
         The Netherlands.

Sector	Energy* (PJ)	Exergy efficiency	PI- index	CO_2 (Mton	CDI- index
Power gen.	492	0.40	43	34.5	52
Industry	846+	0.35	81	57.3	94
Transport	335	0.15	42	24.4	52
Household & Other	755	0.10	100	44.2	100

*Source: [10]
+including feedstocks for the petrochemical industry.

2.2 The size of the CO_2 problem

A crucial factor when searching for options to reduce CO_2 emissions is the huge amount of carbon dioxide produced. Two simple examples will illustrate this fact.

Suppose the tank of a car contains fifty litres of gasoline. The use of that one tank results in a carbon dioxide exhaust of 115 kg. Taking into account the carbon dioxide produced in the refining process, the total amount becomes 127 kg CO_2 ! Thus, by just using one load of gasoline in a car, a person has produced much more than his own weight in terms of carbon dioxide.

The growing of trees and plants results in a net intake of carbon dioxide and an exhaust of oxygen. This respiration varies during night and day, summer and winter, but the effect of a growing season is the storage of carbon in branches and leaves. Due to this respirative action, an increase of forest area is sometimes advocated.

We have investigated the consequences of this proposal for the Dutch situation [11]. What forest area is required to store the annual carbon dioxide production of the Dutch society in its growth of one year? The stunning outcome was that we need 87 times the present forest area or 8 times the total area of The Netherlands. This shows clearly that if reforesting is to contribute considerably to the reduction of the carbon dioxide level in the atmosphere, international co-operation is imperative.

Another way to illustrate the size of the problem is to compare it to the SO_2 and NO_x productions in the use of fuels. Techniques to reduce the emission of these pollutants generally require energy, but this amounts perhaps to a few per cent of the energy involved in the original operation. However, to reduce the carbon dioxide emission with e.g., 90% there is on a short term no other way than to reduce the use of fossil fuels with 90%. Options to do something with the carbon dioxide once it is produced, are discussed in the next paragraph. None of them has short term feasibility.

Conclusion: the major aspect of the carbon dioxide problem concerns the size of the carbon dioxide emission: the amounts involved are roughly a factor 100 larger than those leading to acid rain (sulfur oxide and nitrogen oxides). The only short term option to reduce the carbon dioxide emission is to reduce fossil fuel using productions.

3. OPTIONS FOR EMISSION REDUCTION

3.1 Exergy matching

The need to improve the exergy efficiency of society was stressed already. The PI-index and the CDI-index indicated the sectors with the largest improvement potential. To understand why these sectors have low exergy efficiencies, processes are analyzed by considering the exergy of the supply (donor system) and the demand (acceptor system). The exergy efficiency of a process can be high only when the quality of the demanded energy and the quality of the supplied energy are not too far apart. In that case, a large part of the donated energy can be accepted and conserved. Note that the enthalpy efficiency of a process is always 100%, when all flows, waste or useful, are accounted for. (Conservation of energy principle/First Law of Thermodynamics). Exergy, on the other hand, is lost in irreversible processes, and once lost, cannot be retrieved.

The example mentioned in the previous paragraph, heating of buildings, is typical for a situation of bad match between donated and accepted energy. In fig. 1 the demand of energy (enthalpy) of the Japanese industry is plotted against the temperature and against the quality of the demand. The quality is defined as the ratio of the exergy to the enthalpy. The quality is one for high grade energy such as electricity, mechanical energy and most fossil fuels (a more detailed description of this subject can be found in [12] and [13]).

The mismatch is obvious. The major part of the demand needs an energy source of relatively low quality. On the supply side mainly high quality fossil fuels are used: the overall exergetic efficiency is low. Improving the exergy efficiency of society can contribute considerably to the reduction of the carbon dioxide emission.

A number of improvement options will be discussed in the following paragraphs.

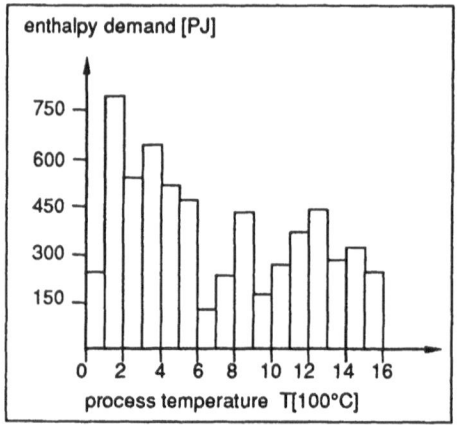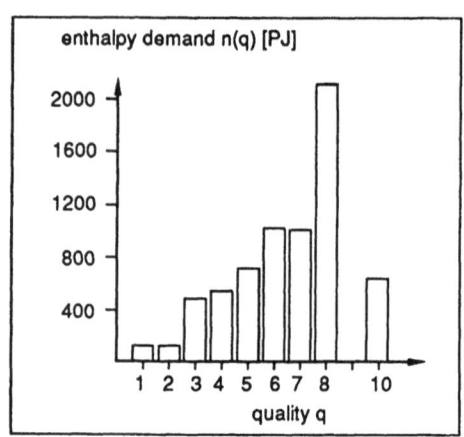

Figure 1. Enthalpy demand of Japanese industry at the different temperature and quality levels [15].

3.2 Power generation

Power generation often receives a great amount of attention when emission reduction of pollutants is considered. For gases such as sulfur dioxide and nitrous oxide, this is understandable since technically sound options exist to remove these components from stack gases. Furthermore, power plants offer the possibility to tackle about 20% of the Dutch energy consumption in a limited number of large units. For carbon dioxide emissions the situation is different. Options for carbon dioxide removal from stack gases are still under investigation. The exergy efficiency of power generation is high, compared to the mean exergy efficiency in other sectors. The major option for improving power generation with respect to carbon dioxide emissions is a shift in fuel from coal, through oil and natural gas towards nuclear power generation and wind turbines.

Future power generation based entirely on wind and nuclear energy would be most beneficial. However, the possibility to apply wind turbines is limited and nuclear power generation is a politically difficult issue. On a short term, the increased input of natural gas in power generation seems the only option. In the present (1987) situation the total carbon dioxide emission from power generation amounts to 34.5 Mton CO_2 per year. Generation of the same amount of electricity based solely on natural gas fired plants would give an annual exhaust of 27.6 Mton CO_2. The total annual emissions in The Netherlands would be reduced by 4.2% (162.5 Mton ---> 155.6 Mton). It will take a number of years before this shift can be completed. In chapter 4 the lifetime of installed capacity and the expected installation of new capacity will be taken into consideration when considering the effect of increased natural gas input in power generation.

3.3 Household & other

The sector Household & Other energy consumers are defined in Dutch Central Bureau of Statistics publications [14]. This sector offers the largest potential for carbon dioxide emission reduction and, at the same time, is one of the most difficult to tackle (together with transport). Large carbon dioxide emissions are generated in small amounts in (central) heating systems in every house, shop, office and building. Therefore, removal of CO_2 from stack gases in this sector is virtually impossible. Improving the exergy efficiency of the heating system is marginal as long as natural gas (or another fossil primary energy carrier) is used: most of the quality of the fuel is lost when low temperature heat is produced. The long term option of improving the insulation of housing (energy use < 1,000 m^3 a.e.) combined with electrical heating through heat pumps is discussed by van der Veer and van Wunnik.

3.4 Industry

According to table 2, industry provides the second largest improvement potential, mainly due to a large fossil fuel consumption. The exergy efficiency in this sector was improved considerably in the last decade. The remaining improvement potential lies mainly in increasing the exergy productivity of smaller plants and in restructuring industry in such a way that full use can be made of heat handling techniques (heat exchange, cogeneration, several types of heat pumps, heat transport and heat storage). The restructuring of industry cannot be done overnight. Considerable progress has been made in the analysis of the options [6,13,15-18]. A computerized version of the exergy analysis and the exergy optimization is currently under development.

3.5 Transport

Options for carbon dioxide emission reduction in the transport sector face the same problem as options for the sector household: a large number of "small" sources is responsible for the emissions. The options to consider here are:
1. Make transport more efficient. Engines are being designed to make better mileage.
2. Use larger units, avoid transport if vehicle is not optimally used. This results in suppressing personal traf-

fic - with very often just one or two persons in a car-
and the stimulation of public transport (buses, trains
etc.). The difficulties this option will encounter became very obvious when the issue of abolishing fixed tax
deductions for commuter traffic led to the fall of the
Dutch government in 1989.
3. Use a different fuel.
a. Electrically driven vehicles.
Especially in urban areas the use of electric vehicles
can contribute to the reduction of CO_2 emissions. Usually travel distances are small and no great speed is required. Moreover, traffic frequently comes to a halt.
Electrically driven vehicles use power only when in mo-
tion, there is no stationary mode. Batteries could be
charged during the night, thereby improving the demand
pattern for electricity, resulting in more (higher efficiency) base load power plants. The largest beneficial
effect can be reached, of course, when non fossil generated electricity is used. Use of fossil fuel power
plants has the advantage that one large power plant is
more suitable for emission reduction measures than
thousands of cars. When this option is realized, air
quality in urban areas will improve greatly.
In table 3, the carbon dioxide exhaust of gasoline fuelled cars and electrically driven cars are compared. For
electrically driven cars, the exhaust takes places at
the power generation site. We have assumed a use of 0.27
kWh per kilometre (0.43 kWh/mile) for electric vehicles
[19] and a 40% power generation efficiency. Gasoline
fuelled cars were assumed to use one litre of fuel for
10 km (0.88 kWh/km).

TABLE 3. CO_2 exhaust for gasoline and electrically
driven vehicles.

		emission factor (g/MJ)	CO_2-exhaust. (g/km)
Electricity	coal	94	226
generated by:	oil	75	180
	nat. gas	56	134
Gasoline	excl.refinery	73	234
	incl.refinery	80	256

On a short term, however, the market for electric vehicles will be relatively small. People will be reluctant
to buy a vehicle that can not serve all purposes
(longer distances). The situation may improve if the
range of the car increases and the infrastructure for

recharging batteries is developed. A possible short term market, besides the use in utility fleets, may be for the second family car. The use pattern of a second car is often more adapted to the specific qualities of an electric vehicle (shorter distances, urban areas, etc.).
b. Use of methanol as motor fuel.
Since methanol is a carbohydrate fuel and the CO_2 emission factor is only slightly lower than that of gasoline (69 kg/GJ for MeOH, 73 kg/GJ for gasoline), something special must be done to obtain an important reduction of carbon dioxide emission with the use of methanol. This will be discussed in paragraph 3.7.

3.6 Removal of CO_2 from stack gases by physical methods

The removal of CO_2 from stack gases by physical methods and the dumping in deep ocean or injection in gas or oil fields is discussed elsewhere at this conference (page).

Application of these options costs a great deal of extra energy. A German study showed that for cryogenic removal of 85% of the CO_2 from stack gases (CO_2 is "frozen out"), subsequent transport to and injection in deep ocean can cost about 30% extra electricity [20]. Cheaper methods appear to be available [21].

3.7 Chemical "capture" of carbon dioxide

We are searching for chemical routes to use carbon dioxide in stack gases from power plants. Exergetically, carbon dioxide is at a very low level. Additional energy input is required to upgrade the CO_2 to e.g., CO, thus facilitating further reactions. Furthermore, the presence of nitrogen in the combustion air leads to an inert bulk in the stack gases. Therefore we have chosen for combustion with oxygen.

The set up we studied for carbon dioxide utilization from stack gases is presented in fig. 2. The air separation unit, the power plant and the reformer section are natural gas fired. Data concerning energy input and CO_2-exhaust are presented in table 4.

Natural gas and oxygen are burned to produce steam for electricity generation and a CO_2/H_2O stack gas. An amount of natural gas is introduced, producing a syngas consisting of CO and H_2. Part of the mixture is recycled (methane, carbon dioxide and steam). We have chosen for external heating to simplify the calculation of the composition of the gases. It is possible to consider autothermal operation (part of the methane feed is burned in the reactor to provide for the necessary energy). This will require extra energy input for air separation, but will lower CO_2

emission in the reformer section. The ratio of CO to H_2 in the syngas is about 0.5. Shifts in composition are possible by introduction of excess steam (more hydrogen) or import CO_2 (more CO) [22,23].

The syngas can be used to produce chemicals. The best application, from the viewpoint of reducing carbon dioxide emissions, is to use the syngas to produce chemicals that are to be "stored" in a commodity with a long lifetime. Production of plastics for furniture is a good example. The carbon dioxide is stored for the lifetime of the product. An important fraction of the capacity of refining industry can be taken over by this production.

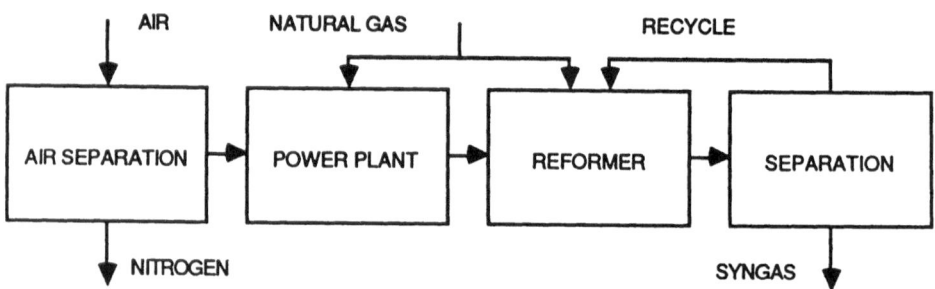

Figure 2. Set up for utilization of power plant stack gases.

TABLE 4. Energy input/CO_2-exhaust when stack gases are used to produce syngas.

	External energy input (GJ)	useful output		CO_2-emission (kg)
Air sep.	0.2	201	kg O_2	11.2
Power gen.	2.5	1	GJ_{e_1}	0
Reformer	7.5	12.5	kmol CO	0
		25.0	kmol H_2	
Heating	2.5			140.0
Total	12.7			151.2

To estimate the potential of this method we have looked into the chemical industry of the United States. We studied the amount of carbon stored in carbon dioxide emitted by power plants and the amount of carbon stored in nine chemicals that rank highly when scaled to production quanta. It appears that the amount of carbon in CO_2 is 2.5 times larger than the amount of carbon in the chemicals. When additional carbon, added to CO_2 to produce the syngas, is also taken into account, then the total amount of carbon obtained from the utility operations is about 16 times the amount of carbon in the top ten carbon containing chemicals. The market for products that can be made out of syngas is much too small to allow the stack gas utilization method to be used to its full potential.

When, however, another large scale application for the syngas can be found, making other productions redundant, an important reduction of carbon dioxide emission might be obtained.

Such an option might emerge in the United States, under impulse of the Clean Air Act [24]. Due to strict regulations on air pollution, especially in urban areas, research is going on to develop methanol fuelled car engines. If this development continues, a large methanol market will develop in the near future.

TABLE 5. Data of methanol production: conventional and stack gas utilization (MeOH-production: 321 kg, power generation: 1 GJ_{e1}).

		external energy input (GJ)	CO_2-emission (kg)
Conventional*			
MeOH	fuel NG	1.6	89.6
	process NG	9.3	
	total	10.9	
Power		2.5	140.0
Total		13.3	229.6
Proposed set-up		12.7	151.2

*Source: [23]

As can be seen from table 5, the proposed method produces power and methanol more efficiently and with smaller carbon dioxide emissions than the conventional production. With relatively small extra investments, part of the syngas can still be used for production of other chemicals. The proposed method and its implications will be discussed in more detail in a subsequent paper [25].

Conclusion: Carbon dioxide emissions are so huge that chemical storage in present day society can only contribute to a very limited extent. Large scale application of chemical storage of CO_2 requires fundamental changes in power generation and chemical industry. A significant contribution cannot be expected on short or medium term.

Summary: based upon exergy matching of demand and supply, six applications are mentioned which might lead to lower carbon dioxide emissions. The options are: shift of fuel in power production, the use of electrical heat pumps to heat homes and buildings, restructuring of industry to improve the exergy efficiency by heat handling techniques, electric vehicles for city transport, physical methods to remove carbon dioxide from stack gases with storage, and the chemical reuse of carbon dioxide from stack gases. The limitations are discussed. No single option has a potential for reducing the carbon dioxide emission on a short term.

4. INTEGRATION OF OPTIONS

4.1 Introduction

Since none of the options has the potential to "solve" the CO_2 problem, a combined effort in several sectors will be needed. The maximum level of application and the time needed to reach this level must be taken into account. These kind of studies are usually investigated by the use of scenarios. A number of possible futures is developed and the effect of the options is investigated. Result of these studies often is that "it can go either way". The outcome varies from positive to negative, usually also predicting a number of intermediate values. If extra attention is given to a certain outcome, this is due to the fact that, to the investigator, this particular result seems more "reasonable" than the others. Atmospheric carbon dioxide levels e.g., or carbon dioxide emission growth rates are presented as results of assumptions about economic growth rates, energy consumption, and the like. The tendency in many of these studies is to focus on the figures produced, rather than the underlying assumptions. We have used a basically different approach. We have chosen 2015 as the year in which to stop the growth in carbon dioxide production. This is not too far into the future to loose sight of reality, but far enough away to allow some measures to get around the inherent delays of society and to be implemented to a significant extent. Figures about penetration of new techniques become variables rather than questionable results. Estimation of an economic growth rate becomes a goal to reach rather than a calculated guess. The result of this kind of simulations is a path to follow to reach a certain objective, not an attempt to foresee the future (see also [26]).

The computer simulation language DYNAMO is used. This language is very suitable for performing time variant simulations. An extensive description of the language is given in [27].

Studying the different sectors we formulated a number of assumptions. As far as they are important for the interpretation of the results, they will be presented here.

Economic development:

As a measure for economic activity the Gross National Product (GNP) is used. We assume a growth of 2% per year in the period from 1989 to 2015.

Electricity demand:

The growth in electricity demand is estimated to be 1.8% per year, the mean value of growth from 1989 to 2008, as can be derived from [28].

Transport:

Emissions from this sector are assumed to keep their present (1989) level, although the number of cars is still growing (The mean annual growth since 1980 is about 2% [29]). This leaves room for improvement due to government measures such as improving public transport.

Household & other:

This sector also is assumed to maintain 1989 emissions. The number of houses is growing steadily by 2% each year since 1975 [30], but there is still a potential for insulation. Twenty-nine per cent of the houses has no insulation at all [31]. In future, old non-insulated houses will be replaced by new better insulated ones, thereby reducing the mean per house emission.

Annually some 115,000 new houses are built. Supposing these are well insulated terraced homes, using about 1,500 m^3 natural gas for heating (the mean use for heating of the present terraced homes in The Netherlands is about 1835 m^3 [31]), an amount of 5.5 PJ has to be saved by insulation measures in existing houses every year, to maintain a constant emission level.

The average use of natural gas for heating purposes in houses is 2,175 m^3 [31]. If these houses could be insulated to use 1,500 m^3 gas for heating purposes only, every year some 257,500 houses have to be insulated! Even if the newly built houses would be insulated to use 1,000 m^3 for heating, still 173,000 houses have to be insulated. This effort has to be continued for as long as the growth in the number of houses lasts. The result is a mere stabilization of carbon dioxide emissions in this sector.

Power generation:

We use the presently installed capacity (incl. 1,800 MW coal fired capacity to be completed by 1995 [28]). Every power plant to be taken into production after 1995 was assumed to be fired by natural gas. It is not commented here whether or not this is a politically sound development, nor whether it should be Dutch gas or import. It is merely chosen as the most likely option to be applied to some extent at least.

Industry:

As was mentioned earlier, most large industrial productions have been optimized already for minimum fuel input. Optimizing smaller industries or large industrial complexes for better exergy efficiencies is much more tedious and less rewarding when CO_2 emissions are considered. Present efficiency aproaches 35%. We assume an annual efficiency growth rate of 1%, starting in 1990 and lasting until 2005. This will result in an overall exergy efficiency of 50%.

4.2 Results

The results of the simulation are plotted in figure 3. The upper line represents the annual carbon dioxide exhaust when no measures are taken. Economic growth and energy consumption continues, no measures in transport or housing, fuel supply as in 1989. The lower line represents the exhaust due to the cumulative effect of the measures presented above. From 1990 to 2005, the period of efficiency improvement in industry, the growth of CO_2-emissions has a constant but significantly lower value than when no measures are taken. From 2005 onward, when efficiency improvement potential in industry has become small, economic growth drives energy consumption and carbon dioxide emissions upward.

Conclusion: The measures proposed above (extensive efficiency improvement in industry, shift towards natural gas fired power plants, maintaining 1989 emission levels in Transport and Household & Other despite a growing number of emission sources) cannot bring about a stabilization of carbon dioxide emissions in 2015. In a situation of growing economic activity more drastic measures will be necessary.

It is also observed (not in fig. 3) that the substitution of fossil fuels in the electricity production by nuclear or wind energy gives only a limited reduction of the CO_2 emission. However, such a substitution is essential to obtain a major reduction of the carbon dioxide emission on other sectors such as Industry, Transport and Household & Other. The options for these sectors rely on an increased use of electricity.

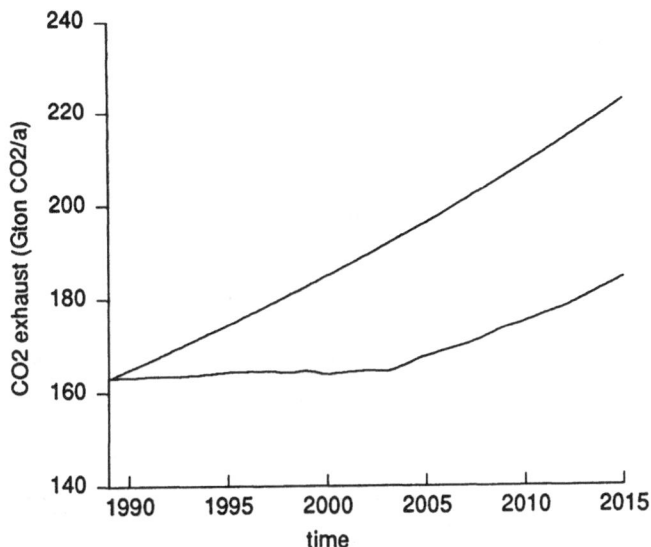

Fig. 3: Carbon dioxide emissions in The Netherlands (Gton CO_2/a);
upper line: no measures, lower line: proposed measures.

5. CONCLUSIONS

We demonstrate the importance of exergy analysis as a tool
to investigate the improvement potential of the energy productivity of society. It is shown that the sectors Household & Other and the sector Industry have the best potential for improvement of both energy productivity and carbon
dioxide emissions.

The major aspect of the carbon dioxide problem concerns the size of the carbon dioxide emission: the amounts
involved are roughly a factor 100 larger than those leading
to acid rain (sulfur oxide and nitrogen oxides). The only
short term option to reduce carbon dioxide emissions is to
reduce the fossil fuel using activities.

Based upon exergy matching of demand and supply, six
applications are mentioned which might lead to lower carbon
dioxide emissions. The options are: shift of fuel in power
production, the use of electrical heat pumps to heat homes
and buildings, restructuring of industry to improve the
exergy efficiency by heat handling techniques, electric
vehicles for city transport, physical methods to remove
carbon dioxide from stack gases with storage, and the chemical reuse of carbon dioxide from stack gases. The limitations are discussed. No single option has a potential for
reducing the carbon dioxide emission on a short term.

A number of measures - extensive efficiency improvement in industry, shift towards natural gas fired power plants, maintaining 1989 emission levels in Transport and Household & Other despite a growing number of emission sources - are collected in a simple model to investigate whether or not they can lead to a stabilization of carbon dioxide emissions by 2015, in a period of growing economic activity. The measures proposed above cannot accomplish such a stabilization. In a situation of growing economic activity more drastic measures will be necessary.

6. REFERENCES

1 Clark, W.C. ed.,"Carbon Dioxide Review: 1982", Oxford University Press, 1982.
2 Hilleman, B., Chemical & Engineering News, 67(11) (1989) pp. 25-44.
3 Keepin, B., Ann. Rev. Energy, 11 (1986) pp. 357-92.
4 Oman, H., in Proc. 21st Intersoc. Energy. Conv. Eng. Conf., San Diego, California, U.S.A, 1986, Am. Chem. Soc., New York, 1986, pp. 66-70.
5 Kram, T., P.A. Okken, Energiespectrum 3 (1989) 66-76.
6 Van Gool, W., F.M.J.A. Diepstraten, "Exergy matching for efficient energy conversion", paper presented on the 2nd Int. Congr. on Energy, Tiberias, Israel, 1988.
7 Diepstraten, F.M.J.A., J. Pruis, E. Nieuwlaar, "Substitution of natural gas by methanol in fertilizer production", paper presented on the 2nd Int. Congr. on Energy, Tiberias, Israel, 1988.
8 Blok, K., S. Fockens, J. Bijlsma, P.A. Okken, "CO_2-emissiefactoren voor brandstoffen in Nederland", Energie Studie Centrum report ES-WR-88-12, Petten, 1988.
9 Marland, G., "The impact of synthetic fuels on global carbon dioxide emissions", in [1], pp.406-10.
10 Netherlands Central Bureau of Statistics, "Energy supply in The Netherlands - Annual figures in 1987".
11 Van Rossum, T., "Het CO_2-probleem", Student report 8907 E, Energy Science Project, State University Utrecht, The Netherlands, 1989.
12 Van Gool, W., Energy, 12(6) (1987) pp. 509-18.
13 Van Gool, W., R. Kümmel, "Limits for cost and energy optimization in macrosystems", in Energy Decisions for the Future, Vol. I, pp. 90-106, Miyata, Matsui eds., Tokyo (1986).
14 [10], pp. 138-9.
15 Kenney, W.F., Energy conservation in the process industries", Academic Press, Orlando, Florida (1984).
16 Kümmel, R., H.M. Groscurth, W. van Gool, "Energy optimization in industrial models", in Proceedings of the 13th IFIP-Conference on System Modelling and Optimization, Iri, M., Yajima, K. eds, pp. 518-29, Springer,

Tokyo (1988).

17 Groscurth, H.M., R. Kümmel, W. van Gool, "Thermodynamic limits to energy optimization", Energy, 14(5) (1989) pp. 241-58.

18 Groscurth, H.M., R. Kümmel, W. van Gool, "Cost aspects of Energy Optimization", in Proceedings of the 14th International Conference on Computing and Information (ICCI '89), Toronto, Canada, 23-27 May, 1989, Koczkodaj, W.W., Janicki, R. eds. Elsevier, Amsterdam (1989), in print.

19 Purcell, G., EPRI-journal, 14(1) (1989) pp. 52-4.

20 Fricke, J., U. Schlüssel,R. Kümmel, "CO_2-Entsorgung", Physik in unserer Zeit, 20 (1989), pp. 56-60.

21 Blok, K., C.A. Hendriks, W.C. Turkenburg, "The role of carbon dioxide removal in the reduction of the greenhouse effect", paper presented at the IEA/OECD expert seminar on energy technologies for reducing emissions of greenhouse gases, Paris, 1989.

22 Moore, R.B., "Economic feasibility of advanced technology for hydrogen production from fossil fuels", Int. J. Hydrogen Energy, vol. 8 (11/12) (1983) pp. 905-11.

23 Goff, S.P., S.I. Wang, Chem. Eng. Prog., August 1987, pp. 46-53.

24 Technieuws Washington, Ministry of Economic Affairs, November 1988.

25 Diepstraten, F.M.J.A., W. van Gool, "The use of flue gas carbon dioxide in chemical processes", in preparation.

26 Landsberg, H.H., Commentary to: Perry, A.M., "Carbon dioxide production", in [1], pp. 364-6.

27 Roberts, N., D. Andersen, R. Deal, M. Garet, W. Shaffer, Introduction to computer simulation, Addison-Wesley Publ. Co., 1983.

28 Electriciteitsplan 1989-1998, N.V. SEP, Arnhem (1989).

29 Netherlands Central Bureau of Statistics, Pocket Yearbook 1988, pp. 250-51.

30 [29], pp. 105-6.

31 Fact and figures of the public gas supply, VEGIN, October, 1988.

THE POTENTIAL OF RENEWABLE ENERGY TO REDUCE CO_2 EMISSIONS

E.H. Lysen
Lysen Consulting Engineer
Pieter Bothlaan 34
Amersfoort
The Netherlands

ABSTRACT. The classic renewable energy sources, biomass and hydropower, prevent a substantial CO_2 emission, by contributing more than 16% of the world energy demand, and their potential for expansion is considerable. Growing more biomass can play an immediate role in absorbing excess CO_2 at a relatively low cost. Of the "new" renewable sources passive solar techniques have great energy conservation potential, wind power is the cheapest source of electricity, and solar PV offers good possibilities to replace a substantial amount of fossil fuels on the long term.

1. INTRODUCTION

One of the key issues of the report "Our Common Future" of the World Commission on Environment and Development is the potential role of renewable energy sources in achieving a sustainable world (1):

> The Commission believes that every effort should be made to develop the potential for renewable energy, which should form the foundation of the global energy structure during the 21st century (p. 195).

In this chapter the present and future contribution of renewable energy sources are discussed with a focus on biomass, hydropower and wind power. The photovoltaic conversion of solar radiation into electricity will only briefly be discussed here, because it is treated in more detail in a separate chapter of this publication. Estimates are made of the contribution of renewable energy sources in reducing the buildup of CO_2 in the atmosphere.

A large variety of renewable energy sources is available, which is both their strength and weakness. Some sources, such as solar and wind power, have a fluctuating supply and are still relatively capital intensive. Their mass application will require further development of proper storage and conversion technologies. Other sources, such as biomass, have a natural CO_2 absorbing capacity, are relatively cheap, but are socially and politically more complex. It is a challenge, though, to build a sustainable world energy structure with this variety of sources, gradually replacing the use of fossil fuels and stabilizing the atmospheric CO_2 concentration.

2. PAST AND POTENTIAL ROLE OF RENEWABLE ENERGY SOURCES

2.1 History

Biomass has been the prime cooking and heating fuel of mankind, and still serves the cooking fires of millions of households today. Biomass directly or indirectly provides our food and has created all our fossil fuel reserves in the past.

Direct solar energy has been used since ages to dry agricultural products, to provide space heat in cold seasons or to create ventilation in homes. The development of water wheels had a strong influence on the expansion of the Roman Empire: horses, previously used to drive corn mills, were freed for military use (2). The use of more than 10,000 windmills made The Netherlands the world's most industrialized country of the 17th century.

Despite the contributions of these other renewable energy sources it was biomass that dominated the energy supply, to such an extent that the depletion of wood usually induced the downfall of a civilization. Britain faced such an energy crisis in the beginning of the 18th century, when iron melting was still heavily dependent on the supply of charcoal and iron works were located near forests, rather than near iron deposits. Coal was used in Britain for hundreds of years, and London even had an antipollution law of 1273, forbidding the burning of coal in the city. But it was the discovery by Darby in 1709, of a method of melting pig-iron using coke, which enabled the industrial change-over to coal. This spurred the development of a whole new range of energy technologies - with profound social consequences - starting with the Newcomen engine designed to pump water from underground coal mines (2).

The necessity to incorporate more renewable energy resources in the world energy structure, as recommended by the Brundtland Commission (1), will similarly stimulate the development of a range of new energy technologies, and may have important social effects.

2.2 Energy flows on earth

To illustrate the potential of renewable energy sources the different energy flows on earth are presented in table 2.1 below. The figure for the world energy consumption, 348 EJ/yr, is the commercial energy consumption in 1988 (11), i.e., excluding the estimated 40 EJ/yr of noncommercial energy, mainly in the form of biomass.

Table 2.1 Flows of solar energy (3, 4, 11)

Solar radiation intercepted by the earth	173,000	TW
Solar radiation reflected by the earth	52,000	TW
Solar energy involved in direct heating	81,000	TW
Solar energy involved in evaporation	40,000	TW
Solar energy utilized in photosynthesis	100	TW
Man's rate of energy use (1988: 388 EJ/yr)	12	TW*)

*) 1 TW (terawatt, 10^{12} watt) corresponds to $10^{12} * 365 * 24 * 3,600 = 31.5*10^{18}$ joule/yr = 31.5 EJ/yr (exajoule/yr).

A conclusion from table 2.1 is that the solar energy flow can be regarded as a very large energy resource, which in principle can supply all energy needs of mankind. For comparison the situation in The Netherlands, with a land area of 33,900 km², is shown in table 2.2 below (5).

Table 2.2 Solar energy and The Netherlands

Average solar energy on Dutch land area	3.6	TW
Energy use in The Netherlands (1987: 2.7 EJ/yr)	0.085	TW

To put these figures in perspective: if solar energy would be converted with an overall efficiency of 20%, the supply of half of the present Dutch energy consumption would require 6% of the land area. For comparison: 8.5% of the Dutch land area is covered with buildings (5).

3. CHARACTERIZATION OF RENEWABLE ENERGY SOURCES

3.1 Types of renewable energy sources

Nearly all renewable energy sources are directly or indirectly derived from the radiation of the sun:

Direct: solar radiation converted into heat (actively and passively),
          electricity or into a chemical substance

Indirect: biomass, hydropower, wind power, wave power, ocean temperature differences

Others: tidal power (attraction of the earth by the moon and the sun)
geothermal power (radio-active decay of isotopes)

In the limited space of this chapter the discussion will be focused on the renewable energy sources which already have a sizeable contribution (biomass and hydro power) and to renewables which have a certain level of technical and economic development and hold a good promise for a reasonable contribution in the near future (wind power, solar heat and solar electricity). This does not imply that geothermal power, for example, cannot have an important local contribution.

3.2 Biomass

The annual production of biomass is of the order of 100 billion tons of dry matter per year on land, and about 50 billion ton/year in the oceans. To compare: the world grain production equals 1.8 billion ton/year and the wood harvest is roughly 1.5 billion ton/year (6). The present consumption of biomass for energy purposes is estimated at 40 EJ/yr (1.4 TW), or 12% of the world consumption of commercial energy.

The potential impact of energy derived from biomass resources is very large, but at the same time biomass is a socially and politically rather complex renewable energy option. Realization of its potential involves resource management (land, water, nutrients, capital), technology development and transfer of technology and the awareness of - and possible adjustment to - social and political constraints (7). One example is the potential conflict with land use for agriculture.

A distinction should be made between two main possibilities to use biomass resources (7): (a) improve the use of existing resources (organic wastes and forest residues) and (b) plant new resources (such as energy plantations). The first possibility is a short-term option, and also involves efforts to reverse the present trend of deforestation. The second option, with respect to the fixation of CO_2, is a typical long term option in view of the large areas required. In this chapter the second option will be discussed.

Role of biomass in the carbon cycle

Biomass, and the biomass of forests in particular, plays an important role in the carbon cycle of the earth. Land clearance of forests for agricultural purposes since the 18th century has been a major contributor (150-180 GtC, equal to 550-660 GtCO$_2$, see note below) to the cumulated man-made release of CO$_2$ into the atmosphere (15). Fossil fuels up to now released 180 GtC[*], i.e. 660 GtCO$_2$. Roughly half of this total amount has been absorbed by the oceans, as described elsewhere in this publication. The present total amount of carbon contained in all biomass on the earth is estimated at 2,000 GtC, of which roughly 1/3 is above ground and 2/3 is in the soil (16).

Disregarding any contribution by burning fossil fuels, there is a large annual flux of about 250 GtCO$_2$/yr (70 GtC/yr) emitted by decaying biomass into the atmosphere, to be absorbed later by newly formed biomass. The result is a permanent fluctuation of the CO$_2$ content in the atmosphere (of about 6 ppm), with a maximum at the end of autumn, and a minimum at the end of spring in the Northern Hemisphere. The relations between the large <u>pool</u> of biomass and the resulting <u>fluxes</u> back and forth to the atmosphere are complicated and not yet fully understood (8). Some authors even argue that the forest/soil complex has been a major sink of CO$_2$ since 1940 (8). It may be clear that the actual role of biomass in the carbon cycle is much more complicated than can be described here. See also the chapter on biomass in this publication.

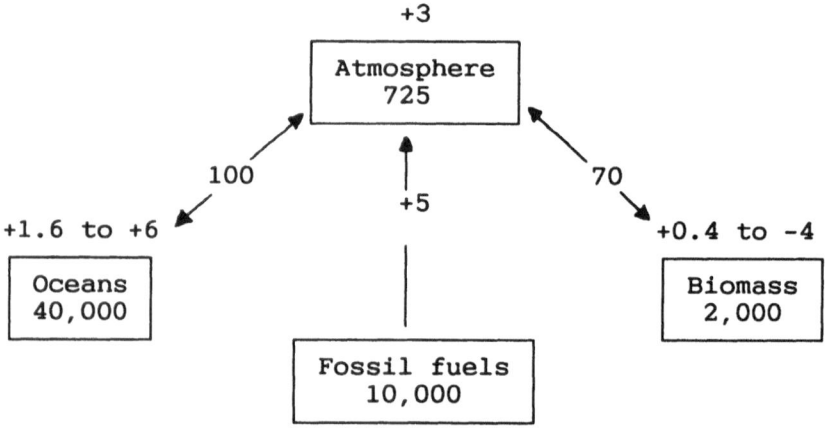

Fig 3.1 Simplified carbon cycle (pools in GtC, fluxes in GtC/yr) (16)

CO$_2$ fixation potential

Every kilogram of (dry) wood contains roughly 0.5 kg of carbon, implying the fixation of about 1.8 kg of CO$_2$

(*) 1 gigaton of carbon (GtC) = 44/12 = 3.67 gigaton of CO$_2$ (GtCO$_2$)

from the atmosphere. The annual production rates of wood strongly depend on species, climate and fertilizer input, and varies from about 4 to more than 60 tons of wood per ha per year (4, 8). This leads to an annual CO_2 fixation rate between about 7 tons CO_2/ha/yr to more than 120 ton CO_2/ha/yr. With a reasonable target of 35 ton CO_2/ha/yr one can calculate that the total area required to absorb the present net CO_2 emission of about 3 GtC/year (= 11 $GtCO_2$/year) by burning fossil fuels, is equal to 3 million km^2, or 2% of the 149 million km2 of land area of the earth. The total forested area of the world, depending on definition, is of the order of 40 to 50 million km^2 (4), which leads to the conclusion that it seems not impossible to realize the additional 7% or 3 million km^2.

An important conclusion from the fact that the present forested area needs to be expanded by about 7% to balance the fossil CO_2 emission is that the CO_2 concentration in the atmosphere must be quite sensitive to small changes in the large fluxes between the standing biomass and the atmosphere, and also the fluxes between the oceans and the atmosphere, both of which are at least an order of magnitude larger than the man-made CO_2 emission.

Realization of potential

It is argued here that it is not necessary to start the required large (re)forestation efforts in developing countries, as is sometimes suggested. This not only because it would suggest that we burden them with solving our problems, but also because there are ample opportunities in Europe and North America, where most of the CO_2 is released.

On the other hand, if developing countries have their own reforestation programmes, such as for example in India and China, and they would require financial and technical support, then it would be unwise not to realize a part of the required forest area in these countries, provided the parties involved can agree on a set of conditions, such as a sustainable use of the forests. An important part of this support could be helping to reverse the present deforestation trend, such as already has been requested by Brasil after international pressure to preserve their tropical rain forests. Several options are available for the required increase of biomass (9):
* Afforestation of areas not (longer) available for agriculture
* Better use of existing commercial forest area
* Increased use of agro-forestry
* Promotion of social and community forestry
* Other types of forestry

Biomass as a renewable energy source

It must be realized that growing more trees represents only a temporary solution for the CO_2 increase of the atmosphere, if not at the same time fossil fuels are gradually replaced by other non-CO_2 emitting sources or by biomass sources themselves. Biomass eventually will decay and release its carbon as CO_2 back into the atmosphere. This may happen in the forest itself after a few decades, or after a longer time span when the wood has been used, e.g. for construction purposes. The wooden poles on which the City of Amsterdam has been built may hold their carbon for two centuries or more, but sooner or later even these poles have to be replaced and their CO_2 is released. This implies that a permanent, i.e. sustainable solution can only be found if the wood in question is not left to decay, but is used as an energy source, thereby avoiding the use of fossil fuels for that purpose. This does not mean that wood is the sole energy carrier, as wood can be converted into the gaseous, liquid and solid forms of energy carriers in use today. They are sometimes called biofuels, to distinguish them from fossil fuels.

In other words: to create a sustainable energy structure, with a stable atmospheric concentration of CO_2, biofuels must gradually replace fossil fuels. This gigantic task is made easier the lower the energy demand (through conservation efforts and population control) and the larger the fraction of the energy demand which is supplied by other renewable energy sources, such as hydro, wind and solar power.

Example: if roughly half of the present world energy consumption, i.e., 175 EJ, has to be supplied by biomass sources, characterized by the production figures mentioned above, then a total area of 6 million km² has to be exploited sustainably.

Costs of CO_2 fixation

Contrary to other renewable energy sources, planting biomass is characterized by low capital costs. Investment levels are in the range between US$ 100 to 600/ha, with US$ 250/ha as a reasonable average (9). If only the fixation of CO_2 is foreseen, i.e. no exploitation of the forest for energy purposes, these are the dominant costs. With a capital recovery factor of 10%, or US$ 25/ha/yr, and a fixation of 35 ton CO_2/ha/yr the CO_2 "removal" costs become US$ 0.71 per ton CO_2.

Costs of biomass for energy purposes

If the biomass has to be used for energy purposes then

the variable cost for maintenance and harvesting tend to dominate. In Sweden much practical experience has been gained with energy forests (9). The average costs of the final product, i.e., wood chips, for a client such as a district heating plant, is 85 Swedish Crowns per thermal MWh (US$ 14 per MWh(th), or US$ 3.9 per GJ, in 1987 prices) and is expected to decrease. A Dutch study for energy plantations in Surinam (1980) resulted in a cost of DFl 6.60 per GJ (US$ 3.3 per GJ), including processing into chips and transport over a maximum distance of 40 km. Grassi (EC, DG XII) mentions the costs of wood in Europe: ECU 30-40/ton in the South and ECU 50-60/ton in the North (10). With 15 GJ/ton and a conversion rate of US$ 1.1/ECU this corresponds to US$ 2.2 - 2.9/GJ resp. US$ 3.7 - 4.4/GJ (1988 prices).

It may be concluded that the lowest wood production prices are in the same range as the present fossil fuel prices, varying in 1988 from US$ 2.1/GJ (coal) to US$ 2.8/GJ (natural gas) in The Netherlands.

3.3 Hydropower

Hydropower is an established renewable energy source, which is derived form solar energy by means of the evaporation of water that is subsequently returned to the earth as rainfall. Mountain sides and streams provide low-cost "solar collectors" and the artificial reservoirs created by large dams can be relatively cheap storage systems. Hydro-electricity tends to be cheaper than electricity derived from either fossil fuels or nuclear power (4).

The theoretical world hydro electric potential has been estimated at more than 44,000 TWh/year, the technically usable amount at more than 19,000 TWh/year and the economic potential at 9,700 TWh/year (4). The largest potential is in the USSR (3,900 TWh), Argentine (2,400 TWh) and in China (1,900 TWh). The economic potential of 9,700 TWh/year is equal to an average power of 1.1 TWe, equivalent to a thermal power of 3.3 TW. The actual world hydro-electric production in 1986, 1987 and 1988 was 2,067, 2,096 and 2,112 TWh/year (11), an average growth of 1% per year.

In the study by Goldemberg et al., (4) it has been assumed that in 2020 the electricity production by hydro electric plants has been doubled to 4,030 TWh/year, a growth of 2% per year. At that time their base case scenario assumes a total electricity demand of 15,600 TWh, of which hydro provides 26%. The 4,030 TWh/year represents 41% of the economic potential, and 21% of the technical potential. This means that it is sufficiently far from the technical limit of the resource that it need not involve sites that would be particularly disruptive ecologically (4).

3.4 Wind power

Although water pumping windmills are an important asset to guarantee the water supply of farmers and villages in many countries of the world, their potential contribution to the world energy demand is small compared to the potential contribution of electricity generating wind turbines, and therefore this paragraph will focus on the latter.

Two trends emerge when reviewing the development of wind turbines the last ten years: (1) investment costs have come down and (2) outputs have increased. As a result the costs per kWh have reduced sharply.

Investment costs

A few years ago the investment costs were in the range of US$ 500 to US$ 1000 per m² of swept rotor area, with a minimum at rotor diameters of around 15 m (power range 50 to 75 kW). At present these figures have been reduced to US$ 300 to US$ 400 per m², and the minimum has moved upward to diameters around 25 m (power about 250 kW) and is still moving upward. Note that these are off-factory figures. The total installed cost of a wind turbine, or a group of wind turbines in a wind plant, strongly depends on local costs such as import taxes, legal fees, roads (if necessary), foundations and grid connection for example. The total cost may increase to 1.5 to 2 times the cost of the wind turbine.

Output

The performance of a wind turbine can be expressed by a so-called "quality factor", which depends on the design of the wind turbine and varies with the annual mean wind speed. Measurements taken at the ECN test station for wind turbines in Petten, The Netherlands, indicate that the quality factor for a standard wind regime of 6 m/s (at 10 m height) has improved from values below 1.5 in 1980 to values above 3.0 at present, i.e., the output per m² of swept area has more than doubled.

Electricity costs

Estimates for the U.K. in 1988 indicate that wind turbines at high wind speed areas would generate electricity at 3 pence per kWh, or US$ 5 cents/kWh (12).

In a recent and detailed study by the Energy Study Centre of ECN in Petten (13) the economic costs of electricity generation are analyzed. Wind turbines of 16 m, 25 m, 34 m and 45 m diameter, with installed powers of 75 kW, 250 kW, 500 kW and 1000 kW respectively, were chosen. The costs in various wind regimes are analyzed, the effect of wind turbines in line or in square parks, and also the effect of lower future turbine costs by applying learning curves.

As an example the results are presented for a line of ten 16 m, 75 kW wind turbines at a coastal site with 6 m/s annual average wind speeds at 10 m height, i.e. 7.1 m/s at the shaft height of 30 m. The total electricity production of the wind plant is estimated at 1.9 million kWh/year. The specific total installed investment cost in 1987 Dutch guilders for a site in The Netherlands was calculated to be Dfl 750/m2 in 1990 (US$ 340/m^2). With a 5% real interest rate and 15 years lifetime the resulting kWh price is Dfl 9.9 cents/kWh (US$ 4.5 cents per kWh). In the same study the wind electricity prices for the year 2000 were estimated, resulting in a minimum for 25 m diameter wind turbines (power 250 kW) of Dfl 7.0 cents/kWh (US$ 3.2 cents/kWh).

3.5 Solar energy

Passive solar energy

Solar energy can considerably reduce the heating and/-or cooling loads of buildings, by applying so-called passive solar techniques. Proper orientation of housing schemes, reducing the size of North facing windows and increasing the South facing windows, attaching sunspaces, increasing the mass of the buildings etc., are all measures which strongly influence the energy consumption for space heating of buildings and houses in temperate and cold climates. Recently the use of transparant insulation materials as passive means of solar heating receives attention. Applying these materials at the outer side of walls increases the heat gain of the walls which is transmitted to the inside of the building. In hot climates the use of passive solar techniques can considerably reduce the cooling load of buildings.

Active solar energy

The active use of solar energy, with residential water heating as the predominant application, is now well established. Solar flat plate collectors for domestic hot water supply can be seen on many houses, not only in the Mediterranean areas but also in Northern Europe. In countries such as The Netherlands a domestic solar hot water system with 3 to 4 m^2 of collector area and a 100 to 150 storage reservoir typically can provide 50% of the water heating requirements of an average family. In the USA, Japan, Israel and Australia several millions m^2 of collectors have been installed. The prices of complete single family solar hot water systems are in the range of US$ 300 to 600 (1984) per m² of collector area (7).

Solar thermal power

By concentrating (direct) solar radiation with mirrors to a receiver, relatively high temperatures can be obtained. The heated steam or oil can be used to drive a mechanical Rankine or Stirling cycle engine which in turn drives a generator. The most successful application commercially used now is the parabolic trough system, in which large rows of parabolic mirrors are used to heat oil, which drives (via steam) a Rankine cycle engine. A total of 160 MWe has been installed, mainly in sunny areas in Southern California, where they can provide peaking power. The unit size per plant is 15 to 30 MWe, and the present average annual efficiency (solar radiation to electricity) is 20% to 25%. Present cost level is around US$ 3,000/kWe. It is to be noted that these systems require a high level of insolation as well as a high fraction of _direct_ radiation to operate properly. That means that they are not well suited for Northern Europe for example, but excellent for the Middle East climate, California, etc.

Solar PV

Solar PV is discussed in more detail elsewhere in this book. It is sufficient here to mention that the efficiencies of solar cells are moving upward steadily (laboratory cells have recorded efficiencies above 31%) and that prices continue to decline. Present module costs are around US$ 3 to 4/Wp, complete systems typically cost between US$ 8 and 10/Wp. Mass fabrication will bring the module costs down to US$ 1/Wp within a few years, and ultimate costs are estimated between US$ 0.2 to 0.4/Wp (19). With an investment of US$ 0.4/Wp, an electricity production between 1 and 2 kWh/yr/Wp (temperate and hot climates) and annual costs equal to 10% of the investment, the electricity costs vary between US$ 2 to 4 cents/kWh. This target cost level will enable solar PV to enter the market of large scale electricity generation.

3.6 Comparison of renewables in energy generation per km²

Some examples to use one km² of land for the renewable generation of electricity or heat, are shown in table 3.1 below. For those sources generating electricity also the _equivalent_ amount of heat to generate electricity in a thermal power station with 33% efficiency is shown. This efficiency is lower than can be attained by present power stations, but is in line with the conversion factor used to represent hydropower in international energy statistics: 1 million ton of oil equivalent (43.2 PJ) is required to generate 4 billion kWh (14.4 PJe) (11).

Table 3.1 Possible usage of 1 km² of land for energy generation   (Note: some applications may be combined)

Subject	Electricity TJe/yr/km²	Heat TJ/yr/km²
1. Incident solar energy:		
Netherlands (horiz.):		3,600
Netherlands (45 degr):		4,200
Sunny site (horiz.):		6,300
Sunny site (angle):		7,000
2. Biomass:		
4 ton biomass/ha/year at 18 GJ/ton		7
20 ton biomass/ha/year at 18 GJ/ton		36
60 ton biomass/ha/year at 18 GJ/ton		108
3. Solar flat plate		
Net system efficiency 40%		
Collector area/land area = 0.10		
Netherlands:		170
Sunny site:		280
4. Solar trough		
Net system efficiency 25%		
Collector area/land area = 0.25		
Installed capacity: 62 MWe/km²		
Sunny site:	440	(1,310)
5. Solar PV:		
Net system efficiency: 15%		
Module area/land area = 0.33		
Installed capacity: 50 MWe/km²		
Netherlands: 58 GWh/yr/km²	209	(627)(eq)
Sunny site: 97 GWh/yr/km²	350	(1,050)(eq)
6. Wind: 32 turbines of 250 kW (8 MW/km²)		
D = 25 m; grid: 5 D * 10 D		
Coast:V(10m)=6m/s 16 GWh/yr/km²	57	(171)(eq)
Land:V(10m)=5.5m/s 12 GWh/yr/km²	43	(129)(eq)

This table may first be used to falsify the popular belief in The Netherlands that the country "has more wind than sun": even at windy sites **one can harvest four to five times more solar electricity than wind electricity in The Netherlands.**

(Note: this statement is not generally applicable, and combining forces is also possible: between the wind turbines rows of solar modules can be installed, increasing the output of the land area).

The next remarkable finding is the prominent place of the solar electric technologies, both solar thermal (trough system) and PV. This is partly due to the fact that the complete land area is filled with solar troughs or panels (the values of .25 or .33 for the ratio of collector (panel) area to land area is a maximum to prevent shading and to provide space for necessary infrastructures), but mainly due to the fact that electricity is being generated, which replaces thermally generated electricity. When solar PV, converted into hydrogen for example, will start to penetrate into direct fossil fuel applications, such as in the transport sector, then lower values apply because in that case the electricity has to be converted into a fuel, e.g., hydrogen (19).

4. CO_2 AND RENEWABLE ENERGY SOURCES

4.1 CO_2 emissions of renewables

During operation most of the renewable energy sources do not emit CO_2, except for open-cycle OTEC (Ocean Thermal Energy Conversion) technology and for geothermal plants, which can be installed at suitable sites (not in The Netherlands). During construction, however, materials and energy are being used, which represent a certain emission of CO_2 as long as fossil fuels are still used. The US Department of Energy has made a comparative analysis of CO_2-emissions of different electricity generating technologies, including renewables (17). For biomass a wood plantation feeding a power plant is assumed, and the net negative value for CO_2 emission is caused by the carbon storage in the roots.

4.2 CO_2 fixation potential

Biomass is the only renewable source which is able to fix CO_2 and therefore takes a special place. As discussed in section 3.2 the costs of CO_2 fixation are modest, and calculated to be roughly US$ 0.7 per ton CO_2. This is based upon an average investment of US$ 250/ha and a CO_2 absorption of 35 ton/ha/year (above ground). US DoE calculations even assume 165 tons CO_2/ha/year, half of which is below ground (17).

Table 4.1 CO_2 emissions electric technologies
(tons CO_2 per GWh)
(NA: Not Applicable; OTEC: open cycle)

Technologies	Fuel Extraction	Construction	Operation	Total
Conventional Coal Plant	1.0	1.0	962.0	964.0
AFBC Plant	1.0	1.0	960.9	962.9
IGCC Electric Plant	1.0	1.0	748.9	750.9
Oil Fired Plant	-	-	726.2	726.2
Gas Fired Plant	-	-	484.0	484.0
Ocean Thermal Energy Conversion	NA	3.7	300.3	304.0
Geothermal Steam	0.3	1.0	55.5	56.8
Small Hydropower	NA	10.0	NA	10.0
Boiling Water Reactor	1.5	1.0	5.3	7.8
Wind Energy	NA	7.4	NA	7.4
Photovoltaics	NA	5.4	NA	5.4
Solar Thermal	NA	3.6	NA	3.6
Large Hydropower	NA	3.1	NA	3.1
Wood (sustainable harvest)	-1509.1	2.9	1346.3	-159.9

(-) Missing or inadequate data for analysis, estimated to contribute ≤1%.

(NA) Not Applicable

*This analysis considered construction of new dams. According to a recent Federal Energy Regulatory Commission report there is 8,000 MW of small hydropower under construction or projected, much of it involving refurbishing or refitting existing dams, which would substa. 'ially reduce small hydropower's CO2 impact.

4.3 CO_2 reduction potential of renewables

On the basis of the production figures in table 3.1 estimates can be made of the CO_2 reduction potential of renewable energy sources. Similar to the energy replacement value mentioned above, the CO_2 reduction values strongly depend on the kind of energy technology which will be replaced. For the electricity generating mix of The Netherlands in the period 1990-2000 an average CO_2 emission of 713 tons CO_2 per GWh has been calculated (18). The average US value for the 1986 fossil-fired production mix was 874 tons CO_2/GWh (17). With the above used power plant efficiency of 33% and an emission factor for coal burning of 94 ton CO_2/TJ (18) a value of 94 * 3.6 * 3 = 1,015 tons CO_2/-GWh results. Taking into account the emission values of table 3.2 and the fact that in a range of developing countries, among which large countries such as India and China, a large coal-fired capacity exists with less efficient power stations, a world-wide replacement value of 1,000 tons CO_2/GWh seems reasonable (the world average is roughly 750 tons CO_2/GWh).

Table 4.2 CO_2 reduction potential of renewable energy sources replacing electricity or heat (see also table 3.1)
Conversion factors: - electricity: 1,000 ton CO_2/GWh
(equal to: 278 ton CO_2/TJe)
- heat (coal): 94 ton CO_2/TJ

Subject	Tons of CO_2/km² /year Electricity	Heat
1. Biomass: 5 ton biomass/ha/year (18 GJ/ton)		846
20 ton biomass/ha/year (18 GJ/ton)		3,384
2. Solar flat plate		
Netherlands: 170 TJ/yr/km²		15,980
Sunny site: 280 TJ/yr/km²		26,320
3. Solar		
Sunny site: 122 GWh/yr/km²	122,000	
3. Solar PV:		
Netherlands: 58 GWh/yr/km²	58,000	
Sunny site: 97 GWh/yr/km²	97,000	
4. Wind:		
Coast:V(10m)=6m/s 16 GWh/yr/km²	16,000	
Land:V(10m)=5.5m/s 12 GWh/yr/km²	12,000	

As expected on the basis of the high output per km² , the solar electric options turn out to be dominant options to reduce CO_2 emissions on the basis of the above area comparison. In terms of costs per kWh, though, biomass and wind power (at windy sites) at present are still much cheaper than these solar options.

5. CONCLUSIONS

The classic renewable energy sources, biomass and hydro power, at present contribute an estimated 40 EJ and 23 EJ (1988) respectively, i.e., more than 16% of the total world energy demand of 388 EJ.
Hydro power alone prevents the emission of 2.1 billion tons of CO_2 per year (2,112 TWh at 1000 ton CO_2 per GWh). Its potential for expansion is considerable: when the hydro power output is doubled from the present 2,112 TWh to 4,000 TWh this represents still only 40% of the economic potential.
Biomass has the unique advantage that by planting more biomass CO_2 is absorbed, thereby compensating the emission of CO_2 and reducing the increase of the atmospheric CO_2

concentration. With an average fixation value of 35 ton CO_2 per ha per year and an investment of US\$ 250 per ha, the costs are around US\$ 0.7 per ton CO_2, which is rather low. Compensation of all excess CO_2 produced at present would require 3 million km² of land, which equals 2% of the land area of the earth.

Of the "new" renewable energy sources passive solar energy techniques have great energy conservation potential, by reducing the space heating or cooling loads of buildings. For electricity generation wind energy is the cheapest renewable source at present, given a proper wind regime. At a coastal site of The Netherlands the costs are estimated at US\$ 0.05/kWh (1990), decreasing to below US\$ 0.04/kWh in the coming decade. The costs of PV generated electricity is at least five times higher at present, but the rate of investment in new and large PV manufacturing plants is such that the kWh costs decrease rapidly. The ultimate costs, in the beginning of the next century, are estimated to decrease to US\$ 0.02 to 0.04/kWh.

Analyzing the output of the renewable energy sources per km² of land area, one concludes that the solar-electric options yield substantially higher outputs than wind power. Even in a country which is not particularly sunny, such as The Netherlands, solar PV yields four to five times more electricity than wind power per km². Given the fact that the world supply of solar energy is roughly two orders of magnitude larger than wind energy, solar PV on the long term offers good possibilities to replace a substantial amount of fossil fuels and reduce the CO_2 emissions accordingly. This will first be in the form of PV electricity and later in the next century as PV generated fuels such as hydrogen.

REFERENCES

1.   Our Common Future (1987) The World Commission on Environment and Development, Oxford, Oxford University Press.
2.   Foley, Gerald. (1976). The Energy Question, Penguin Books.
3.   Hubbert, M. King (1971). The Energy resources of the Earth, reprinted in "Scientific Technology and Social Change", Readings from Scientific American, January 1974, pp. 263-272.
4.   Goldemberg, J., T.B. Johansson, A.K.N. Reddy, and R.H. Williams. (1988). Energy for a Sustainable World, Wiley Eastern Ltd, New Delhi.
5.   Statistisch Zakboek (Statistical Pocketbook, 1988) Staatsuitgeverij 's-Gravenhage.
6.   Van der Toorn. (1988). Strategy for biomass conver-

sion, Shell International Petroleum Company, Group Public Affairs, London.

7. Renewable Sources of Energy (1987). International Energy Agency (IEA), Paris.

8. Wiersum, K.F. (1989). Personal communication, 16 March 1989.

9. Daey Ouwens, C., and E.H. Lysen. (1989). Herbebossing ter compensatie van de emissie van kooldioxyde veroorzaakt door het gebruik van fossiele brandstoffen (Reforestation to compensate fossil CO_2 emission) Netherlands Ministry for the Environment, Project ELMI.

10. Grassi, G. (1988). Bio-energy-industrial integrated regional projects in the European Community, Proceedings of the Euroforum - New Energies Congress, Saarbrucken, Vol 1, pp. 55-57.

11. BP Statistical Review of World Energy (1989). BP International Limited, London.

12. Lindley, D. (1988). The commercialisation of wind energy, Proceedings of the Euroforum - New Energies Congress, Saarbrucken, Vol 2, pp. 47-70.

13. Van Wees, F. and G. Bakema. (1988). Economische Rentabiliteit Windenergiesystemen (Economic Rentability Wind Energy Systems), Netherlands Energy Research Center, Report ESC-47, ECN, Petten.

14. Daey Ouwens, C. (1976). Does solar energy require more land surface and more materials or energy investment than nuclear energy or fossil fuels?, International Conference on Solar Electricity, Toulouse, France.

15. Trabalka, J.R. (Editor). (1985). Atmospheric Carbon Dioxide and the Global Carbon Cycle, US DoE, Office of Energy Research, Washington.

16. Okken, P. (1987). Energie en het broeikaseffect (Energy and the Greenhouse Effect), Netherlands Energy Research Center, Report ESC- 40, ECN, Petten.

17. San Martin, R.L. (1989). Environmental emissions from energy technology systems: the total fuel cycle. IEA/OECD Expert Seminar on Energy Technologies for reducing emissions of greenhouse gases, Paris, 12-14 April 1989.

18. Okken, P. (1989). Personal communication, 15 June 1989, citing from ECN reports ESC-WR-89-4, and ESC-WR-88-12.

19. Ogden, J.M., and R.H. Williams. (1989). New prospects for solar hydrogen energy: implications of advances in thin-film solar cell technology. IEA/OECD Expert Seminar on Energy Technologies for reducing emissions of greenhouse gases, Paris, 12-14 April 1989.

THE PROSPECTS OF PHOTOVOLTAIC SOLAR ENERGY CONVERSION

W.C. TURKENBURG, E.A. ALSEMA and K. BLOK
Department of Science, Technology and Society
University of Utrecht
Oude Gracht 320
NL-3511 PL Utrecht
The Netherlands

ABSTRACT. Solar energy can be converted directly into electricity by means of solar cells. The costs of these cells have been reduced more than a factor of 20 in the past 15 years. Meanwhile the conversion efficiency has been increased more than expected, with a record value for a concentrator cell of 31%. Further improvements and cost reductions are expected, not only of cells but also of the solar cell modules and solar cell systems. At the moment PV-system costs amount to $ 10 per Watt and less. A cost reduction to below $ 1 - 2 per Watt is expected, although it probably will take two decades.

The market for PV now consists of small scale and stand-alone applications, in developing countries but also in the western world. At the end of the 1990s PV power is expected to be an attractive option for peak loads of the public grid. At costs below $ 1 - 2 per Watt PV power can be a competitive option for base load applications connected to the grid.

In the development of PV much attention is given to thin-layer solar cells like amorphous silicon solar cells. However, the present market is still dominated by crystalline silicon. The market might grow from 35 MW in 1988 to multithousand MW a year in the next century.

1. INTRODUCTION

Solar energy can be used to generate electricity in a direct way by means of the photovoltaic (PV) effect, first recorded by Edmund Becquerel in 1839. Solar cells using crystalline silicon have first been made in the 1950s and are applied to power satellites in space since 1958 [1].

After the publication of the Report to the Club of

Rome, but especially after the 1973 oil crisis, PV systems are considered for terrestrial applications as well. Special R&D programs have been initiated with the objective of developing low-cost PV-systems for isolated, stand-alone applications as well as grid-connected applications. As a result, the selling price of solar cells has dropped from over $ 100 per Watt in the beginning of the 1970s to less than $ 5 per Watt in 1988, whereas in sunny regions the cost of PV electricity has dropped from $ 15 to approximately $ 0.3 per kilowatt-hour [1,2].

Photovoltaic power plants require no fuel and little maintenance. The direct conversion of sunlight-to-electricity is done without any moving parts and causes hardly any environmental damage. The use of PV to generate power is certainly one of the options to reduce the threats of the greenhouse effect caused by the increased use of fossil fuels.

This paper will review the state of the art of PV technology with an emphasis on solar cells. Prospects on cost reductions and its impacts on market development will also be reviewed.

2. SOLAR IRRADATION AND THE CONSUMPTION OF ENERGY

In sunny regions the mean value of irradiated solar energy on the earth's surface amounts to 4.5 - 6 kilowatt-hour per m^2 per day. In the Netherlands this value is 2.75 - 3.00, resulting in a yearly irradiation of about 1000 kWh per m^2. On an inclined surface this value can be higher, up to - in the Netherlands - 17% at an inclination of 45° to the south.

The total amount of energy irradiated by the sun on earth equals almost 10.000 times the commercial energy consumption of the world. For the Netherlands this ratio is about 50, whereas compared with the electricity consumption this value is more than 500.

Knowing that the direct conversion of sunlight-to-electricity by PV can be done with an efficiency of about 10% - and in the future maybe even 20% or more - one might conclude that the world energy need can be supplied in principle by solar electricity. For the Netherlands the situation is more complicated, because of the high energy consumption rate per m^2 and the intensive utilization of land. Assessment studies indicate that on houses and building (roofs, walls) it might be possible to install a PV generating capacity of 50.000 Megawatt, assuming a conversion efficiency of the system of 20% [3]. Such a system might be able to produce 50 Terrawatt-hour per year, about 70% of the electricity consumption we are facing today.

It is concluded that solar PV can be a substantial

source of energy, even on higher latitudes. The major drawback at the moment is the high costs of PV if compared to conventional power production. Worldwide a huge PV R&D program has been set up to overcome this problem.

When grid-connected PV systems have been implemented substantially, a further expansion of PV power requires the availability of cheap energy storage systems.

3. SOLAR CELLS

The basic element of a PV system is the solar cell made from semiconductor material. Semiconductors absorb sunlight by ejecting electrons from the chemical bonds between atoms (see for example refs. 1 and 4). See also figure 1. In crystalline silicon an amount of energy of at least 1.1 eV (accordingly a wavelenght of light of 1.13 µm) is required to create a free electron.

A solar cell has two or more specially prepared layers, for example a p- and a n-type layer, to get a junction that creates an electric field. The free electrons, generated by photons from the sunlight, are driven to the n-type layer. If outside the solar cell the n-type layer is connected to the p-type layer via an electric resistance, an electric current is obtained and the solar cell provides power.

Using normal sunlight, the conversion efficiency of a single crystal silicon solar cell can be, in theory, about 25%. The highest value obtained up to now in a laboratory situation is about 23%. Theoretically, the conversion efficiency of a gallium arsenide (GaAs) cell could be 28%. For amorphous silicon (a-Si) this value is 24%, for copper-indium diselenide combined with cadmium sulfide ($CuInSe_2$/CdS) 22% and for cadmium telluride combined with cadmium sulfide (CdTe/CdS) 28%. In the development of solar cells these materials attract almost all the attention. Recently, however, also organic materials are investigated; although the solar cell research on these materials is in an initial stage the first results look promising.

To get even higher efficiencies than the values mentioned before, two options are explored. The first one is a 100- to 1000-fold concentration of sunlight on the solar cell. This requires the direct irradation to be a large fraction of the total insolation, because diffuse sunlight cannot be concentrated effectively. In the Netherlands only 40% of the solar irradation is direct. Therefore concentration doesn't make sense at moderate climates. The other options is to combine two or three solar cells having different band gaps, each making use of different parts of the solar spectrum.

Along these lines it should be possible to produce solar

cells that convert solar energy to electricity with an efficiency as high as 35%. By applying both options in one system, recently a record efficiency of 31% has been obtained under a 300- to 500-fold concentration.

Not only the efficiency that can be reached is important, but also the thickness of the cell that is needed to trap and absorb the light. In the case of crystalline silicon the thickness of the cell should be at least 100 μm, although new concepts are developed of polycrystalline silicon cells that are much thinner [1]. In the case of amorphous silicon, which is a highly light-absorbing material, only 1 to 2 μm is needed to absorb 99% of the incident light above the 1.7 eV band gap of the material. Thin solar cells can also be made if one uses GaAs, CuInSe$_2$ or CdTe.

Because of the high costs of semiconductor material, the possibilities of thin film solar cells get much attention. Another reason for this interest is, that thin film solar cells and thin film solar cell modules can be produced by completely different and in principle less expensive processing techniques.

Up to now crystalline silicon has been the leading material in the production of solar cells for the power market. Modules with an efficiency of about 10 - 13% are sold at a cost of about $ 5 - 6 per Watt. Per m^2 this means a cost of $ 500 - 800. It is generally believed that, if the necessary investments are made, modules of thick polysilicon solar cells having an efficiency of 17-18% can be manufactured at a production cost of $ 1 - 1.5 per Watt, so $ 170 - 270 per m^2.

The most important alternative for crystalline silicon is amorphous silicon. The first a-Si solar cell was made in 1973 and had an efficiency of 1%. Now, in the laboratory an efficiency of 12% is reached for single-junction cells and of 14% for multi-junction devices. Commercially available modules have efficiencies of 6% or more. One can buy these modules at a cost of $ 4 per Watt, so a cost per m^2 of about $ 250. According to Carlson [4] and other authors [5,6] these costs might fall to $ 0.5 or even $ 0.3 per Watt at a module efficiency of about 10%, indicating that a cost level of $ 50 per m^2 might be reached. This explains why nowadays much attention is focused on amorphous silicon. However, amorphous silicon has two major drawbacks. The first one is the initial loss of efficiency after exposure to light. A lot of research is done to understand this phenomenon. As a result one is able to limit the degradation to about 10% for cells and 15% for modules [1]. The other drawback is the fact that the efficiency is still rather low. It might well be that high cell efficiencies (above 20 - 24%) can only be obtained if one could make an appropriate three-junction (in stead of a one- or a two-junction) device, each having a different band gap. The research that is going on in this field might result in the production of inexpensive modules with efficiencies of 15-20% [1,4,7].

As CdTe and CuInSe$_2$ are investigated still on a relatively limited scale, no definite prospects can be given yet on costs and efficiencies. A record efficiency of about 12% has been reached in the case of CdTe and of about 14% in the case of CuInSe$_2$. It is expected that the production costs of electricity generated with these types of solar cells will be equivalent to the production costs that are obtained with amorphous silicon solar cells.

The highest efficiencies may be obtained with gallium arsenide and its alloys, such as aluminium gallium arsenide (AlGaAs). The record efficiency of a GaAs cell obtained in the laboratory is about 24%. Modules have been made for space applications with an efficiency of about 16% and at an estimated cost of $ 500 per Watt; this equals a value of about $ 80.000 per m^2 [7]. Also in this case cost reductions will take place. However, from this figure it is clear that GaAs cells face the obstacle of costs even more than other types of solar cells.

The progress that the conversion efficiency has made in recent years is depicted in figure 2 for different types of solar cells.

4. PV POWER GENERATION

To be economically viable, the cost of PV electricity must be competitive with the cost of electricity generated from other sources. The cost-competitiveness of PV we are facing at the moment is shown in figure 3. The figure indicates that PV is an attractive energy source if used in small scale applications, like pocket calculators, and in small scale stand alone systems, like telecommunication systems, buoys, street lights and systems to generate electricity for isolated dwellings.

More and more PV becomes an attractive option for larger stand alone power systems, as an alternative to (or in combination with) a gasoline or diesel generator. As the availability of a cheap and reliable energy source in areas not served by the grid is a must for the further development of these areas, PV might play an essential role in this development, provided that the investment costs of PV are reduced further. For that purpose attention should be given not only to the costs of the solar modules, but also to the costs of the other components of the solar system and to the installation costs. These additional costs are indicated with the expression BOS (Balance-Of-System) costs.

It appears likely that PV will make its most signifi-

cant impact on the energy supply of the world if it becomes
an attractive option to generate base load power to the
utilities. This means that the production costs have to be
reduced about a factor of 5 in sunny regions, like California, USA, and about a factor of 10 on higher latitudes,
like the Netherlands.

If the space available on rooftops can be utilized for
the solar arrays, Alsema et al. [3,6] have calculated that
for the Netherlands cost-effectiveness is reached if a reduction of the module cost down to $ 55 per m² at a module
efficiency of 18% is obtained. ᴬt a module efficiency of
30% the module cost that is allowed amounts to $ 120 per
m². In this calculation the following assumptions have been
made:

Break even kWh costs	0.05 $
Mean annual irradation	1170 kWh/m²
Irradation losses	6 %
Power related BOS costs	150 $/kW
Area related BOS costs	20 $/m²
Additional investment costs	25 %
Yearly exploitation costs	1.5 $/m²
Real interest rate	4 %
Economic life	30 years
Average inverter efficiency	95 %

A graphical representation of the calculated results is
given in figure 4, taken from ref. [7]. For comparison,
figure 4 also shows the US DoE goal (in constant 1986 dollars), based on a 30-year levelised cost of 0.06 $/kWh, as
quoted in their Five Year Research Plan for 1987-1991 [12].
It is in good agreement with the estimate for the Netherlands, bearing in mind that it is based on a higher mean
annual radiation (2400 kWh/m² and above) and higher area-
related BOS costs (for multi-Megawatt power plants 50 -
100 $/m²).

It is important to note that, at least in the USA,
before the ultimate target of competitiveness with base
load electricity is reached, photovoltaics will compete
with small (25 MW) conventional peaking generators at a
levelised cost of about $ 0.12 per kWh. It is expected
that this goal will be reached in the second half of the
1990s. Therefore, preparations are made to build a 50 MW PV
power plant near to San Diego. This plant might be realised
within a few years. The construction of the plant will be
based on the use of amorphous silicon. The plant is jointly
developed by the Chronar Corporation and SeaWest Power Systems [1]. The world's largest PV power plant that has been
built up to now, can be found at Carissa Plains, California, and has a generating capacity of 6.5 MW. Experience
with the Carissa Plains facility has shown that solar elec-

tric generators are highly reliable and easy to install and operate [1,13].

Figure 5, taken from ref. [3], shows how the cost reduction of PV electricity might take place [3]. The figure indicates that it will take at least another 20 years before cost-effectiveness of grid-connected PV systems might be expected for the Netherlands.

In this comparison the social and environmental costs, which are avoided with PV solar energy conversion, are ignored. A recent study on the social and environmental costs of energy consumption [8] suggests that these costs are much higher in the case of nuclear energy and fossil fuels than in the case of PV. Therefore, from the perspective of our society, the break-even costs of PV are higher than the values indicated before. Following the calculations of Hohmeyer [8] the influence of the social costs of conventional power generation on the break-even cost of PV is depicted in figure 5 as well.

We might conclude that, starting in the USA, within 10 years a new era in the development of PV will begin in which competitive, central utility power generation is possible.

As indicated, the application of PV power highly depends on the associated costs. Due to cost-reductions the market has grown from 3.3 MW in 1980 to 35 MW in 1988. The growth in 1988 amounted to 23% [9]. Amorphous silicon accounted for nearly 40% of the modules shipped in 1988, while the single crystal silicon cells dropped to 37% of the total shipments. Polycrystal silicon cells stayed at a level of 22% [9]. The leading position in these shipments is taken by Japan (13.0 MW in 1988), followed by the USA (11.3 MW in 1988) and Europe (6.9 MW in 1988).

Worldwide more than 60 industrial companies produce solar cells and modules. In 1988 the largest share in the module shipments has been delivered by ARCO Solar (5.5 MW), followed by Sanyo (4.8 MW), Solarex (3.2 MW) and Chronar (2.9 MW), see [9].

The leading position of Japan and the USA over Europe might change, now the budgets for PV R&D have increased dramatically in Germany and Italy. All together we estimate that the worldwide government expenditures in 1988 amounted to 200 million dollars. Adding the expenditures of the industry, the yearly total investment in PV R&D can be estimated to be 500 million dollars. The share of Germany in these expenditures is about one quarter, the share of the Netherlands about 0.5%.

Several studies have been made to estimate the market potential of PV as a function of cost, see for example refs. [3] and [10]. From these studies one might expect a shipment of 1000 MW/year at a cost of $ 2 - 3 per Watt and as much as 20.000 MW/year at a cost of $ 1 - 2 per Watt.

These figures should be compared with the shipment of 35 MW in 1988 at a PV-system cost of about 10 $/Watt.

Assuming a production level of 20.000 MW/yr and a 30 year lifetime for the PV system, a PV generating capacity of 600.000 MW could be sustained. Such a capacity could produce approx. 1000 TWh of electricity per year, depending on the geographical distribution of the solar systems.

5. FINAL CONCLUSION

Following H.M. Hubbard, the director of the Solar Energy Research Institute, Golden, USA, the final conclusion of this review can be, that PV solar energy has a bright future, not only from economic considerations but also from the consideration of global warming [1]. However, the rate of penetration of PV power systems as described in this paper, can only be realized if a strong and continued R&D effort in this field is supplied.

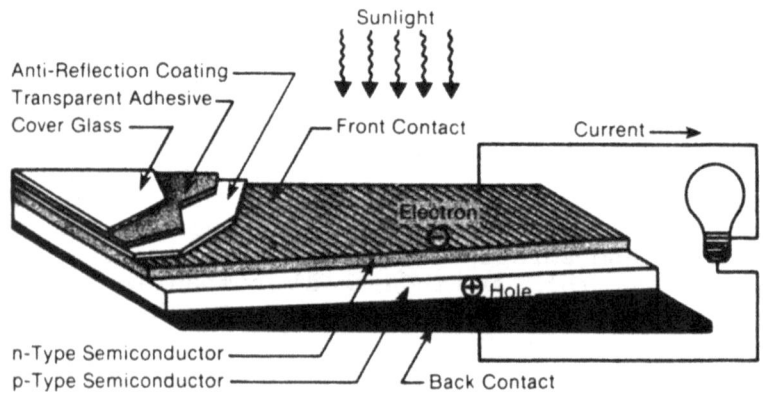

Figure 1. A schematic representation of a PV cell under solar insolation [12].

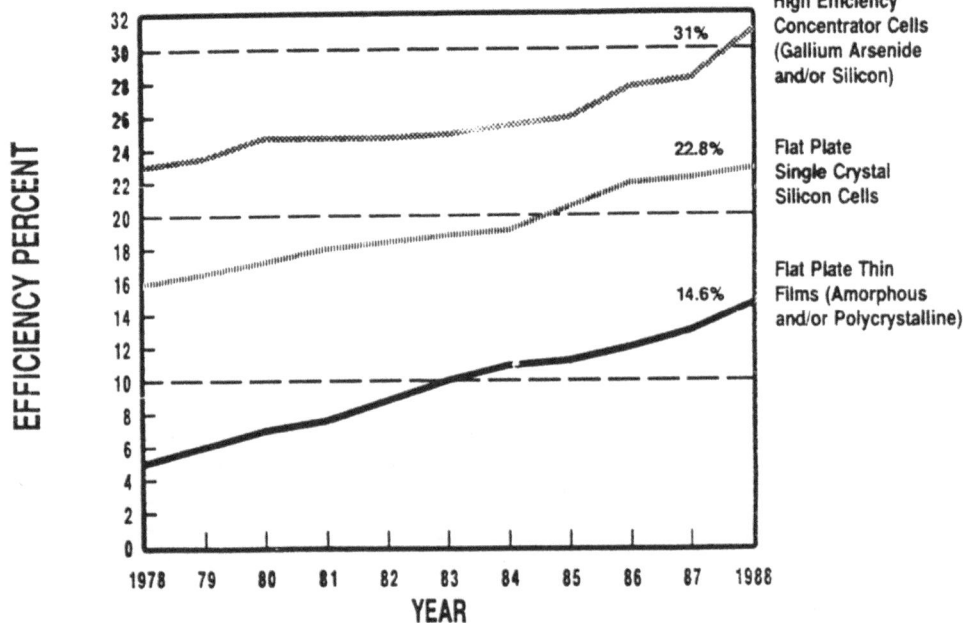

Figure 2. The development of solar cell efficiencies obtained in laboratory situation for the different types of solar cells [11].

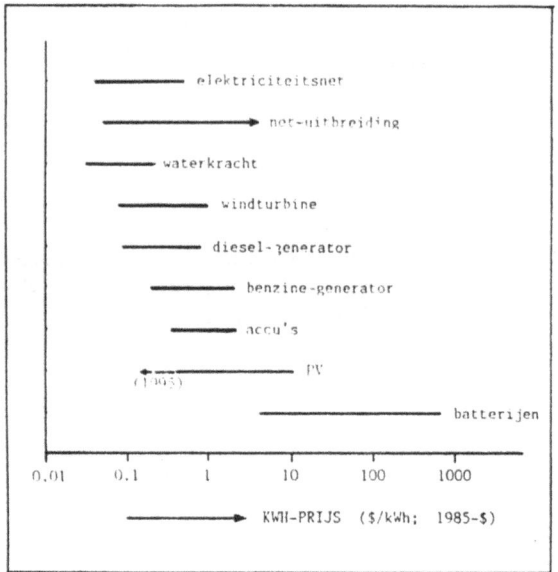

Figure 3. The price of electricity from PV systems compared with electricity prices from other electricity supply systems [3].

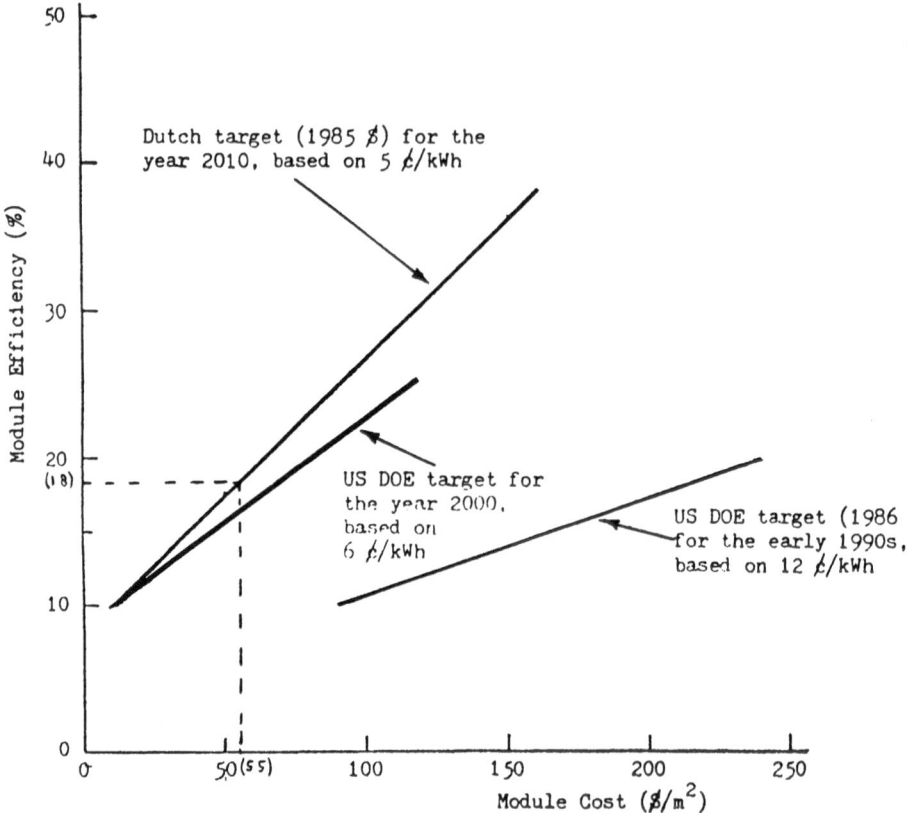

Figure 4. Cost/efficiency targets for modules in grid-connected PV-systems in the Netherlands for the year 2010. For
comparison US DoE's targets are also shown [7].

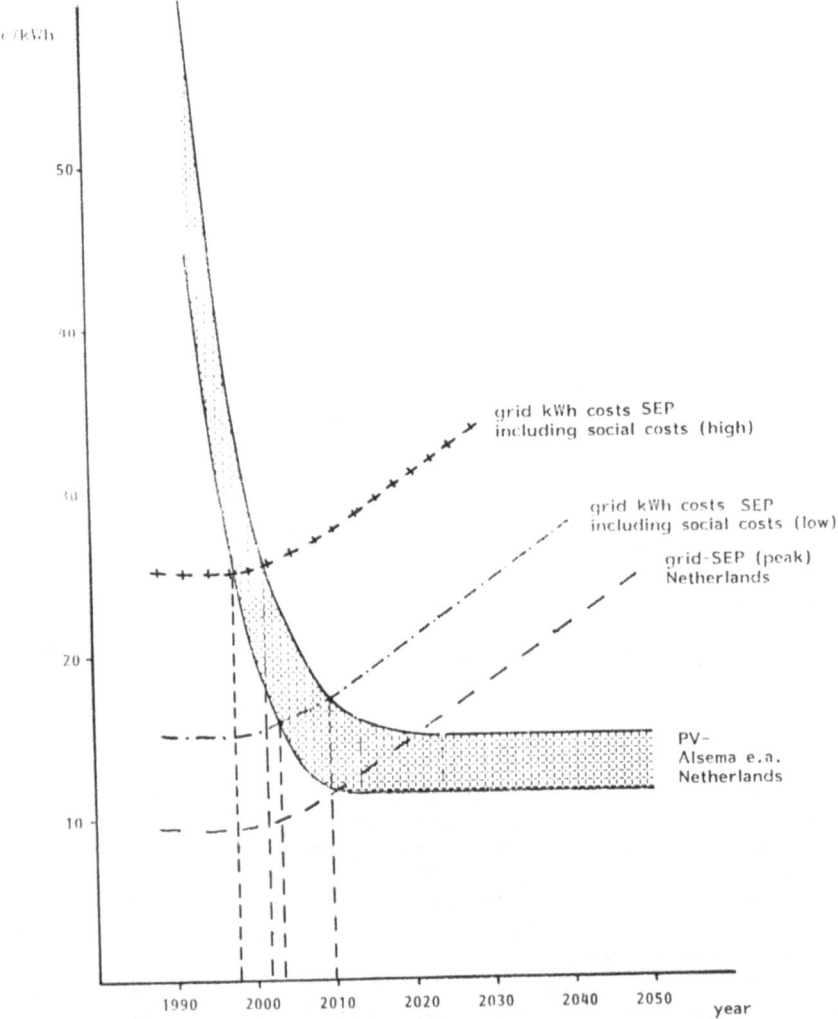

Figure 5. The cost of PV electricity compared to the costs of conventional electricity production with and without the inclusion of social costs [14]. With 'SEP' the Dutch utilities are indicated.

REFERENCES

1. H.M. Hubbard: "Photovoltaics Today and Tomorrow", Science 244, 4902 (April 21, 1989) 297-304.
2. W.F. van der Weg en W.C. Turkenburg: "Solar Photovoltaics; how far are we?", Proceedings of the Second Duth Solar Energy Conference, Noordwijkerhout, April 18-19, 1989 (in dutch).
3. E.A. Alsema and W.C. Turkenburg: "Electricity in the Netherlands with solar cells", Publication Milieubeheer 88-1, Ministry of Environmental Affairs, Leidschendam, 1988 (in dutch).
   E.A. Alsema, K. Blok, E.W. ter Horst and W.C. Turkenburg: "The application of solar cells in the supply of electricity to the Netherlands up to the year 2050", University of Utrecht, February 1987 (in dutch).
4. D.E. Carlson: "Low-Cost Power From Thin-Film Photovoltaics", published in "Electricity", eds. T.B. Johansson, B. Bodlund and R.H. Williams, Lund University Press, 1989, pp. 595-626.
5. J.M. Ogden and R.H. Williams: "New prospects for solar hydrogen energy: implications of advances in thin-film solar cell technology", IEA/OECD Expert Seminar on Energy Technologies to Reduce Emissions of Greenhouse Gases, Paris, April 12-14, 1989.
6. I.F. Garner: "Solar Power Stations For Rural Electrification", Electricity International (january 1989) 53-55.
7. F.C. Treble, E.A. Alsema and W.C. Turkenburg: "Assessment of Dutch Research on Amorphous Silicon and III-V Solar Cells", Netherlands Agency For Energy and Environment (NOVEM), Utrecht, spring 1989.
8. O. Hohmeyer: "Social Costs of Energy Consumption", Springer Verlag, Berlin, 1988.
9. P.D. Maycock and E.N. Stirewalt: "World PV Module Production Up 23 Percent; U.S. Share 30 Percent", PV-News 8,2 (February 1989) 1-7.
10. M.R. Starr and W. Palz, "Photovoltaic Power For Europe; An Assessment Study", D. Reidel Publ. Comp., Dordracht, 1983.
11. This figure has been distributed at the 20th IEEE Photovoltaic Specialists Conference, Las Vegas, September 26-30, 1988.
12. US Department of Energy: "Photovoltaic Five Year Research Plan 1987-1991", Washington, 1987.
13. R.W. Smock: "Photovoltaic solar cells gets a utility tryout in California", Power Engineering (May 1989) 9.
14. R.F.A. Cuelenaere, A.J.M. van Wijk and W.C. Turkenburg: "Social Costs of Electricity Production and the Introduction of Grid Connected PV Systems", University of Utrecht, December 1988 (in Dutch).

REFORESTATION, A FEASIBLE CONTRIBUTION TO REDUCING THE AT-
MOSPHERIC CARBON DIOXIDE CONTENT?

K.F. WIERSUM
Department of Forestry
Wageningen Agricultural University
P.O. Box 342
6700 AH Wageningen
The Netherlands

and

P. KETNER
Department of Vegetation Science, Plant Ecology and
Weed Science
Wageningen Agricultural University
Bornsesteeg 69
6708 PD Wageningen
The Netherlands

ABSTRACT. Deforestation and reforestation cause respectively a release or accumulation of carbon to or from the
atmosphere. Various approaches exist to estimate the
amounts of carbon involved in these processes. Three options for using forestry as a means of CO_2 control are prevention of deforestation, long-rotation plantation forestry, and fuel wood production as substitute to fossil fuel.
Reforestation is also needed for several productive and
environmental purposes; it is estimated that around 200
million ha of new plantations should be established. These
plantations would annually accumulate around 20% of the net
annual increase in atmospheric CO_2. In addition high-yielding fuel wood plantations managed on a recycling basis may
substitute a yearly fossil fuel emission of 0.12 Gt C.

1. INTRODUCTION

The world's forests covering around 30% of the earth's
surface play an important role in the global carbon cycle.
They are a reservoir for around 75% of the total amount of
carbon estimated at 560 Gt stored in the terrestrial biomass. In addition, forest soils store about 55% of the total soil carbon estimated at 1,400 Gt. Combined the forest

ecosystems thus account for a reservoir of over 1,200 Gt C; this is higher than the total amount of carbon stored in the atmosphere (700 Gt C). In addition to forming an important reservoir of carbon, forests also play an important role in its cycling. Through their photosynthetic activity forests can fix atmospheric carbon dioxide. Annually, an estimated 60 Gt C is fixed as net primary production by all world phytomass, most of it by forests. But at the same time CO_2 is released from the biosphere by heterotrophic respiration processes such as the decomposition of dead biomass (1, 2, 3).

During the last decade the role of forests in the global carbon cycle has received increasing attention, particular in relation to the question whether forests form a source or sink for atmospheric CO_2 (4, 5, 6). On the one hand, growing concern has arisen about the possible effects of the emissions of CO_2 through land clearing and forest burning on the increase of atmospheric CO_2 concentration. But on the other hand, it has also been suggested that reforestation may be a possible way to fix part of the increasing atmospheric CO_2 content caused by fossil fuel burning, which gives a net increase of atmospheric CO_2 of ca. 3 Gt C year^{-1}.

This paper will assess whether reforestation can be instrumental in reducing the atmospheric carbon dioxide content, and whether its application is practically feasible. For this purpose, first the effects of deforestation and other land-use changes on the global carbon budget will be reviewed. Secondly, it will be discussed what the possible contribution of reforestation can be to accumulate carbon from the atmosphere and/or to reduce fossil fuel emissions by producing wood as fuel on a recycling basis. The possible contribution of reforestation to carbon accumulation will then be compared with the needs for reforestation for other purposes. Finally, the practical feasibility of reforestation will be evaluated on the basis of recent experience with ongoing reforestation programmes.

2. EFFECT OF DEFORESTATION AND OTHER LAND-USE CHANGES ON THE GLOBAL CARBON CYCLE

During the last decade much discussion has focused on the contribution of deforestation and other land-use changes on the increase in atmospheric carbon concentration. First estimates in the second half of the 1970s indicated that net transfers in the order of 2.6-6 Gt C might result from deforestation (16, 17); a number of estimates even suggested that the annual release from forest clearing could be as large as two to four times those from fossil fuels (7). Since then many efforts have been made to actua-

lize the data on the effects of human activities in the
land biota on the global carbon cycle.

Table 1 gives some of the estimates of the annual flux
of carbon between atmosphere and terrestrial ecosystems. A
more elaborate list is given by Houghton et al. (8). There
are several difficulties involved in estimating these
fluxes. The first difficulty is the lack of precise data on
the global deforestation rate and on the nature of various
forms of vegetation regrowth following forest clearing. An
associated problem is the paucity of reliable data about
biomass values of various vegetation types. A second difficulty arises from the lack of precise information on the
fate of carbon after forest cutting and burning. Seiler and
Crutzen (9) were the first to indicate that a significant
amount of biomass may not be combusted to CO_2 but rather be
transformed to elemental carbon due to incomplete combustion to charcoal. Thirdly there is much uncertainty about
the amounts of soil carbon tied up in the litter and humus,
which are released through increased oxidation after clearing. Finally, it is still uncertain how large the fertilization effect of increased atmospheric CO_2 concentration
on terrestrial ecosystems is. Such an effect is well-proven
for horticultural cultivation in greenhouses, but its effect in natural ecosystems under conditions of limiting
water and nutrients is still open to discussion.

In response to these difficulties, various approaches
to estimating the role of terrestrial biota in the CO_2 budget have been developed (10, 11). First, there are bookkeeping models which evaluate amounts of CO_2 release on the
basis of land-use changes such as deforestation and forest
volumes. Originally, much attention focused on properly
estimating the amounts of destroyed biomass, but gradually
more sophisticated assessments have been made which include
estimates on the carbon sequestered by the recovery vegetation following forest clearing. The most widely quoted recent results of this bookkeeping approach indicate net annual biospheric release rates of 1.0-1.8 Gt C (10, 12).
However, major points of discussion are the possibility
that as yet undisturbed tropical forests, which are considered to be in a carbon steady state, are in reality still
slowly sequestering carbon (13), and the degree to which
forest plantations, especially in temperate zones, are ser-
ving as a net carbon sink because of changes in use and
management as well as expansion in area. The estimates for
temperate forest areas variously indicate a net annual release of 0.1 Gt C (12) to a net sink of 1-1.2 Gt C (4).

The second approach to estimating terrestrial carbon
release consists of the development of dynamic models of
the global carbon cycle. In such models consideration is
given to dynamic changes in assimilation and loss fluxes
which result from deforestation and land-use conversion. It

allows the estimation of the gradual carbon accumulation in young, actively growing vegetation types, and of the effect of growth stimulation through increased CO_2. The outcome of this approach may be time series of annual CO_2 fluxes. Two recent results of such an approach by Goudriaan and Ketner (2) and Esser (14) suggest that the terrestrial biosphere has become a small net sink of CO_2 since around 1970.

TABLE 1. Comparison of estimates of the annual transfer of carbon, circa 1980, from terrestrial biota to the atmosphere, as a result of deforestation and other human disturbance (Gt of C).

	Present release caused by deforestation	Present net release including increased net primary production
Bookkeeping models		
Bolin (15)	0.4 - 1.6	
Woodwell et al. (16)	4 - 8	
Hampicke (17)	1.5 - 4.5	2.5
Seiler & Crutzen (9)	-2 - 2	
Houghton et al. (18)	1.8 - 4.7	
Houghton et al. (12)	1.0 - 2.6	
Detwiler and Hall (10)	0.4 - 1.6[+]	
Dynamic biosphere models		
Goudriaan and Ketner (2)	7	- 0.5 - 0
Esser (14)	2.7	- 0.1
Geochemical models		
Peng et al. (19)		1.2
Emanuel et al. (20)		1.8
Peng (21)		- 0.8

[+] Tropical forests only

The third approach is based on geochemical models. This method tries to evaluate the carbon release of the biosphere by comparing the time history of the atmospheric CO_2 content with the fairly well-known history of the carbon release due to fossil fuel combustion. Originally the time history of atmospheric CO_2 variations was assessed on the basis of $^{13}C/^{12}C$ ratio in tree rings (19, 20). Recently, data on historic trends in atmospheric CO_2 variation was also reconstructed from air bubbles trapped in polar ice. Using these data, Peng (21) concluded that the forest and soil reservoir became a sink for carbon beginning approximately in 1940. By 1980, the uptake by the forest-soil carbon pool reached 0.8 Gt C.

Obviously, the various estimates about CO_2 flux between the biosphere and the atmosphere show large discrepancies. However, the range in estimates has been decreasing and several of the early high estimates have been revised downwards (e.g. 10). The recent model studies have revealed that although carbon accumulation through regrowth may be significant, it does not compensate for the losses through forest burning and soil oxidation. If the land biota are a sink for carbon or in near equilibrium, this is probably due to the fertilization effect of increased atmospheric CO_2 concentrations. These higher concentrations result in higher production values and consequently increased accumulation of carbon in living biomass and the soil. Goudriaan & Ketner's model (2) indicates that the formation of elemental carbon after burning could be a significant carbon sink as well.

3. STRATEGIES FOR REDUCING ATMOSPHERIC CARBON DIOXIDE CONTENT BY FORESTS

Because of the role of forests in the global carbon cycle, it has been suggested that reforestation may be used as a means to reducing the atmospheric CO_2 content. In this respect two different options present themselves. Forest plantations may be either managed to optimally accumulate carbon, or they may be managed to produce wood fuel for use as an alternative to fossil fuel. Furthermore, it should be realized that reforestation will not have any effect, if the established plantations fall prey to the same processes of destruction, which are now causing the disappearance of natural forests.

Consequently, three strategies may be distinguished for using forestry as a means of contributing to the control of the atmospheric CO_2 concentration:
- Prevention of deforestation and forest degradation;
- Plantation forestry for long-term carbon accumulation;
- Wood as substitution for fossil fuel.
These strategies do not necessarily exclude each other, but the kind of forest management practices involved have different characteris-tics. For analytical and planning reasons it is therefore useful to make such a distinction.

3.1. Prevention of deforestation and forest degradation

As discussed, various estimates indicate that due to the large areas of vegetation regrowth including forest plantations and increased atmospheric CO_2 concentrations, at present the land biota may well be acting as a small sink of CO_2. However, this semi-balance may again be disturbed, if tropical deforestation continues at the present rate of about 11 million hectares per year. Any further clearings will significantly influence the carbon cycle as a whole by increasing emissions to the atmosphere, and by reducing resident time of carbon in phytomass and soil, and thus repressing the fertilization effect (14).

Furthermore, it is important to realize the differences in young and mature forests in respect to the carbon budget. Young, actively growing forests are effectively accumulating carbon and thus can contribute to reduce the atmospheric CO_2 content. Mature forests are considered in an equilibrium with a balance between carbon accumulation and release. Thus, they do not effect the atmospheric CO_2 concentration. However, in these mature forests large amounts of carbon are tied up and thus they form an effective carbon reservoir. It takes at least 30-50 years before an equivalent amount of biomass is built up in new plantations. Therefore, it is more efficient to conserve this reservoir, rather than using new forests to sequester similar amounts of carbon. Any effect at influencing the atmospheric CO_2 content through forestry, should therefore start by preventing continuation of the present deforestation process.

Goudriaan & Ketner (2) illustrate that changes in deforestation rates have significant effect on atmospheric carbon. According to their dynamic model, if the deforestation rates had been only half the present ones, the atmospheric CO_2 concentration would have been 10.6 Gt C lower in 1980. And Postel and Heise (22) state that halving the CO_2 contribution from deforestation in the four countries where deforestation rates are highest, would cut total net carbon release from tropical countries (which they estimate at 1.6 Gt (10, 12)) by more than 20%. Thus, any reduction of deforestation would contribute to controlling the atmospheric CO_2 content.

An important factor to be considered in this regard is the hazard of decreasing vitality and production capacity of forests caused by the combined effects of various forms of atmospheric pollution. It does not make sense to propose that forestry should contribute to decreasing the atmospheric CO_2 concentration, if other forms of atmospheric pollution in industrial regions are causing a decrease in forest vitality.

3.2. Plantation forestry for long-term carbon accumulation

From the discussion on the effect of deforestation on the global carbon budget, it is clear that young forests are already actively accumulating carbon. Hampicke (17)

estimated that existing reforestation areas yearly accumulate 0.3 Gt C from the atmosphere, while according to
Houghton et al. (18) secondary vegetations and plantations
account for an annual carbon fixation of 1.3 Gt. Other estimates indicate that tropical and temperate plantations
are a carbon sink of 0.03-0.11 Gt y^{-1} (6) and 1-1.2 Gt y^{-1}
(4) respectively. The variation in these estimates indicate
that the exact effect of existing plantations is still un-
clear. This is caused by the lack of precise data about
reforestation areas (around 88 and 11 million ha in temperate and tropical regions respectively (27)) with their age
distribution and their rate of production. It should also
be considered that the final net effect of carbon fixation
after reforestation partly depends on the nature of the
prior vegetation. For instance, reforestation on grasslands
results in less gains in carbon storage than reforestation
on abandoned cropping lands due to loss of the grass biomass. Similarly, an initial carbon loss may occur in case
of reforestation of shrub or fallow lands.

With the present reforestation technology it is possible to obtain a maximum mean wood biomass production of
up to 10-14 ton $ha^{-1} year^{-1}$ (6, 23), provided that the right
choice of fast-growing species is made and that the plantations are established on reasonable fertile lands in areas
with adequate rainfall and temperature conditions. Such
production would equal a yearly carbon fixation of 4,500-
6,300 kg per hectare. In addition, carbon would be accumu-
lated in roots and the soil; these amounts are not further
considered in the following discussion. Myers (cited in 24)
mentions that 200 million hectares of new forests would
store around 1 Gt C per year, while Postel and Heise (22)
estimate that 120 million hectares of fast-growing tropical
forests would yearly sequester 0.8 Gt C. Thus the realization of 100 million hectares of new plantations would result in the fixation of maximally 20% of the annual increase of atmospheric carbon content of around 3 Gt. However, for long term purposes and over such large areas
(with average growing conditions) such high production values are unrealistic, and much lower values should be considered.

For optimal carbon accumulation, it is necessary that
this carbon is stored for long periods. This means that a
major factor to be considered in carrying out such a strategy for carbon fixation is the necessity for these plantations to grow as long as possible; usually the rotations of
such plantations should be at least 50 years, but preferably much longer. Although the net growth of old forest
plantations is generally lower than that of young forests,
the total amount of carbon stored in old forests is much
higher than in young forests. Long rotations also allow for
optimizing management practices to obtain the highest pos-

sible biomass of trees and the associated undergrowth species, and of humus in the soil.

An interesting question which has hardly been addressed yet, is what the optimal mix would be between an exten-
sion in plantation area and an increase in rotation length.
In this respect it should also be considered, that the effectiveness of reforestation as well as of natural forest
conservation is closely related to effective forest maintenance. If the basic causes of deforestation are not at-
tacked simultaneously, reforestation efforts will not be
effective. Within a few years the plantations too will have
disappeared. This is demonstrated in many reforestation
schemes where hardly any trees remain after a few years
(25).

A further prerequisite for optimal carbon accumulation
is that the wood produced in the new forest plantations is
conserved as long as possible, also after cutting of the
forests. This implies that this wood should be utilized in
long-lasting structures, rather than being burned or decomposed. In fact long-lasting structures such as construction
material, furniture and to a certain extent also books
might be a non-neglectible sink for carbon (26).

3.3. Wood as substitution for fossil fuel

A specific characteristic of forests is that they may
not only be used as a carbon reservoir, but that the wood
produced in forests may also be used as a form of energy.
As the increase of atmospheric CO_2 concentration is mainly
due to fossil fuel combustion, it is therefore logical to
comtemplate whether fossil fuels might be substituted by
wood fuel. Theoretically, in such energy plantations the
annual productivity would equal the amount of wood utilized
as fuel each year. Consequently, a substitution of fossil
fuels to wood fuel would entail a change from a form of
energy utilization which releases carbon dioxide into the
atmosphere to a form of energy utilization which is based
on a closed carbon cycle. The primary aim of such a strategy is thus to decrease the flux of carbon dioxide into
the atmosphere. In addition, such plantations would also
sequester a certain amount of carbon. But as the aim of
these plantations is optimal energy production rather than
optimal carbon fixation, the plantations will offer only
relatively little carbon-fixing benefit, since the carbon
they accumulate during growth will be quickly released
again when burned.

For optimal wood fuel biomass production generally
very short production cycles of 3-10 years will be used in
high-density plantations and with high input levels including fertilization (23). As wood is a bulky product with a
relatively low energy content per weight unit, the trans-

port of fuel wood usually becomes too expensive after a distance of 50-100 km. Consequently, to be a feasible alternative to fossil fuels, such energy plantations should be concentrated near the intended locations of use. Because of this and the fact that economic feasible levels of energy production can only be obtained on good quality soils, usually the plantations will have to compete with other forms of land use. Therefore, this strategy offers most scope for implementation in areas where land is taken out of farming. It is estimated that within the European Community over the next decade over 15 million hectares of land will have to be taken out of farming as a result of the Common Agricultural Policy (23), while in the United States some 16 million hectares of cropland may be taken out of production as conservation reserve under the Food Security Act (22). If these areas were to be transformed into high-productive wood energy plantations with an average rotation of 8 years and average annual wood production equalling fixation of 4.5 ton carbon per hectare, after 8 years a condition will be reached, whereby annually 4 million ha of fuel wood can be harvested. This would replace a fossil fuel flux of 0.12 Gt C year^{-1}. By rejuvenating the harvested areas fuel wood could be produced on a recycling basis. The total carbon pool in the phytomass of these plantations would amount to about 0.8 Gt C.

4. OTHER NEEDS FOR REFORESTATION

The possibility for carbon fixation is but one of several reasons for reforestation. Since the middle of the 1960s there has been a gradually increasing recognition of the urgent need to increase the rate of reforestation. In 1967 the World Symposium on man-made forests was organized as a first effort to discuss how best to tackle the goal of doubling the area of man-made wood plantations by 1985. Since that time, concern about the need for reforestation has only increased. Around 1980 the global area of forest plantations was 99 million hectares, of which 11.3 million hectares were located in the tropics. At that time around 9.2 million hectares were planted annually of which 1.1 million in tropical countries. It was then recommended that the rate of planting in tropical countries be increased five-fold (27).

The call for these reforestation efforts was not based on the desire of reducing the atmospheric CO_2 content. Indeed, there are many other reasons why reforestation should be increased (22, 27), such as the need for increased fuel wood production for the local population in tropical countries (28), the need for increased production of industrial wood (29) and the need for replenishment of degraded tropical lands which are subject to environmental deterioration (30).

In respect to fuel wood FAO (28) estimated that the area needed to cover the present deficits in fuel wood production by rural people in the tropical regions amounts to 48 million hectares, but this may increase to 105 million hectares in 2000. Other estimates arrive at an area of 55 million hectares, provided that 25% of demand is met through fuel substitution and the use of more efficient wood burning technology (22). These estimates concern the need of fuel wood by people who have either no physical or financial access to fossil fuel, but do not include the possibility of substitution of fossil fuels by wood energy as discussed earlier. The fuel wood produced in these plantations will be utilized either for meeting formerly unfulfilled energy demands, or as substitution of other forms of bio-energy such as cow dung or agricultural residues.

Increased rates of reforestation are also needed to produce sufficient amounts of industrial timber. FAO (29) has estimated that to prevent industrial wood deficits in 2000, it will be necessary to replant 20-40 million hectares. For the period 1987-1991 it was estimated that some 10 million hectares of industrial wood plantations should be established in tropical countries (31).

Grainger (30) has calculated that there is at present a total area of 758 million hectares of degraded tropical lands with potential for forest replenishment. Of this area 203 million hectares consist of degraded forest fallow vegetation, 137 million hectares of logged-over forests, 331 million hectares of drylands subject to desertification and 87 million hectares of deforested upland watersheds. Other estimates suggest that 160 million hectares of tropical uplands have become severely degraded and need rehabilitation (31). Although measures for ecological rehabilitation do not only entail tree planting (e.g. ecological stabilization may in many cases also be achieved through mechanical soil conservation measures or grass planting) obviously there is a great need to increase the rate of tree planting for soil and water conservation. Not all of these tree plantations do necessarily have to take the form of closed plantations. For instance, for wind protection in desertified areas trees may be planted as shelter-belts, while for controlling water erosion trees may be planted along contour lines.

In Table 2 the needs for reforestation are given in amounts of equivalent areas of closed plantations. Areas of logged forests and tree fallow vegetation needing replenishment with more economic valuable tree species are only partial included in this table, as these forms of forestation mostly involve merely a change in tree composition rather than an increase in forest areas. But a part of

these lands has been wasted to such an extent, that the
secondary forest vegetation has been replaced by a deflected vegetation of very low stature, e.g. alang-alang (Im-
perata sp) grasslands. Such wasted fallow lands needing
revegetation are estimated to cover 23-40 million hectares
(30).

TABLE 2. Estimates on needs for reforestation (million
hectares of equivalent closed plantations).

Areas needed to cover deficits in wood production	
Fuelwood deficits in 2000	55 - 100
Industrial wood deficits in 2000	20 - 40
Degraded areas needing protective tree cover	
Degraded upland watersheds[1]	44 - 80
Desertified (semi-)arid lands[2]	+ 50
Wasted fallow lands	23 - 40

[1] Assumed 50% tree cover
[2] Assumed 15% tree cover

Source: 22, 27, 30.

5. FEASIBILITY OF USING REFORESTATION AS A MEANS FOR CARBON
   FIXATION

     Forest plantations may fulfil a variety of roles for
human society, producing either wood, energy or food and
fodder products, or providing environmental service functions in the form of water and soil conservation or micro-
climate control. However, it is unrealistic to expect a
forest plantation to fulfil all possible roles simultaneously. Nonetheless, it will be possible that plantations
combine several of these roles; by applying proper multi-
purpose forest management techniques it may be possible to
assure the realization of a combination of productive and
service functions. For instance, reforestation of severely
degraded hill sites will assist in reducing erosion, improving soil qualities and producing fuel wood or fodder for
the local population, but the growth of such plantations
will mostly be too slow for economic viable production of
commercial timber. Alternatively, the cultivation of short-
rotation high-yielding biomass plantations will normally
provide little option for soil amelioration; on the contrary, these plantations will need to be fertilized to sustain
production. Fortunately, reforestation may take several
forms varying from long to short-rotation plantations and
line or scattered plantings. Trees may be either planted as
closed tree plantations or in various mixtures with agricultural crops (agroforestry). This flexibility in forest

management allows for optimizing reforestation designs in response to specific objectives (22, 27, 31).

In order to assess the feasibility of using reforestation as a practical means for carbon fixation, the first question which arises is to what extent it is possible to combine the objective of carbon fixation with other objectives for reforestation, and what kind of plantation designs should be preferred for meeting these combined objectives. Obviously, each succesful establishment of new tree plantations will result in at least temporary fixation of carbon. However, in case the biomass of such plantations is quickly used again for fuel wood or fodder, fixation will be relatively small and its effect short-lived. And in case forest plantations replace natural woody vegetation with low economic value then the overall gain in carbon fixation will only be small or may even be negative.

The best results in sequestering atmospheric carbon are reached in long rotation forest plantations with fast-growing species. Such characteristics are best displayed in industrial timber plantations and to a somewhat lesser extent also in plantations with environ-mental protection objectives. There is a world-wide need for an estimated 127-190 million hectares of new environmental protection plantations and for 20-40 million hectares of new industrial wood plantations (Table 2). If it is assumed that 50% of the industrial wood needs can be obtained from plantations which also serve environmental protection purposes (mainly on degraded watersheds and wasted fallow lands), this would mean that roughly 160 million hectares of new forest plantations will also have good carbon accumulation characteristics. If the average growth rate of such plantations is estimated at 3,000 kg C ha^{-1}year^{-1}, these plantations would be able to fix in the order of 0.5 Gt C annually for at least the next 30 years. This is around 16% of the net annual atmospheric CO_2 increase.

In addition to these plantations further carbon fixation may take place in local fuel wood plantations mainly for subsistence use, of which between 55-100 million hectares are needed (Table 2). If it is again assumed that 50% of the needed fuel wood can be obtained from the environmental plantations, there remains roughly a need for 40 million fuel wood plantings. Because most of these plantations will need to be located in semi-arid or mountaineous areas the average yields of these plantations may be estimated at roughly 2,000 kg C ha^{-1}year^{-1}. This would mean a total annual carbon accumulation of around 0.1 Gt C. This net accumulation would continu over a much shorter period than that in the environmental plantations, because the produced wood will soon be used as fuel.

The above estimates are summarized in Table 3. They should only be taken as indicative values of the order of magnitude of the possible effects on the atmospheric carbon content in case all presently anticipated needs for new forest plantations are fulfilled. The estimates indicate that by the time this total area is realized, this will result only in a reduction of the net increase in atmospheric CO_2 concentration. This area of reforestation would need to be increased by a factor 4 to 5 in order to accumulate the total amount of CO_2 annually released by fossil fuel combustion.

It should be borne in mind that all these calculations are based on static bookkeeping principles, only taking into account changes in the forest phytomass component, whereby no attention is paid to changes in soil carbon. The real situation will be much more complex, because any change in the pools and fluxes of the biosphere will automatically give rise to a chain of alteration in the total carbon cycle, in which the oceans also form an important link. Only by the use of refined dynamic models will it be possible to gain more detailed insights into the actual net changes and effects of reforestation.

Could such large areas of reforestation in reality be realized?

TABLE 3 Summary of estimated effects of various forms of reforestation on carbon accumulation.

	Long-rotation	Fuel wood (mainly subsistence)	Substitution fossil fuel
Area (mil.ha)	160	40	30[1]
Productivity ($kg\ C\ ha^{-1}yr^{-1}$)	3,000	2,000	4,500
Effective period (yr)	30	5-10	*
Carbon accumulation ($Gt\ C\ yr^{-1}$)	0.5	0.1	*
Lowering C emission from fossil fuel ($Gt\ C\ yr^{-1}$)	n.a.	n.a.	0.12

[1] In European Community and USA
* 8 year rotation, total amount of stored carbon 0.8 Gt C

As mentioned in 1980 the area of annual reforestation was around 9.2 million hectares, of which 1.1 million hectares in tropical countries (27). Since the 1970s the rate of reforestation has increased, especially in tropical countries (32); according to WRI (33) the present world-wide rate is 14.5 million. Since 1985 a strong stimulus to tropical reforestation as well as forest conservation is provided by the Tropical Forestry Action Plan (31, 34). But even if the five-fold increase in tropical reforestation, which has been internationally called for (27, 31) should be realized, it would still take decades to establish the 200 million hectares of new plantations needed in tropical countries. This would mean a tremendous effort, because many local ecological and socio-economic factors have to be taken into account for succesful reforestation (27, 32), and no blueprint approach is possible. Of special importance is the participation of the local people in such schemes. Such participation will only be possible in case reforestation designs are used, which address the needs of the local people (22, 27, 35). Numerous examples of failed reforestation projects illustrate, that it is unrealistic to expect rural people to contribute to programmes for long-term benefits, if their own immediate problems are not solved at the same time. This means, that especially in tropical countries where local people are faced with many pressing problems in meeting their daily livelihood needs, the objective of carbon fixation should never be an aim in itself for reforestation. Rather, such carbon fixation should be considered as a welcome by-product of reforestation schemes, whose primary objectives are of a different nature.

Any additional reforestation effort beyond those to cover the needs for wood products and local environmental services can only be expected in areas, where sufficient amounts of land are available which are not needed for subsistence or commercial agricultural production. Such areas are more likely to be found in the industrialized areas with excess in agricultural production than in the tropical regions. Indeed, as most fossil fuel use is taking place in these industrialized countries, it is logical that the problems of atmospheric CO_2 increase are also solved in these countries and that these problems are not transferred to tropical countries.

6. CONCLUSION

Mature forests form an important reservoir of carbon. They are in a steady-state condition with a zero balance between carbon fixation and release. Such forests cannot contribute to the reduction of the atmospheric CO_2 concentration, but conservation of these forests will prevent the carbon stored in them to be released into the atmosphere. Young forests are still increasing their biomass and these forests sequester carbon from the air. However, if the wood produced in these forests is cut and subsequently burned or decomposed, such carbon accumulation is only of a short-term nature. Optimal carbon fixation is thus obtained in plantations with a long rotation. Fuel wood plantations, however, may contribute to reducing the increase in atmospheric CO_2 concentration, if the wood is used as a substitute for fossil fuels.

In addition to the need for carbon fixation, there are many other reasons why reforestation is needed; they relate to the productive and environmental service roles of forests. Forest plantations may provide several functions at the same time, and CO_2 accumulation can be an important side effect of reforestation for other purposes. It is estimated that world-wide there is need for establishing 160 million hectares of new forest plantations, which can fulfil such multiple roles. In addition around 40 million hectares of tropical fuel wood plantations for meeting unfulfilled local energy demands are needed. The establishment of these areas of new forests would result in an estimated accumulation of 0.6 Gt C, which is 20% of the net annual increase in atmospheric CO_2. In addition, the establishment of around 30 million hectares of plantations in the EC and USA for wood fuel production on a recycling basis would result in a decrease of fossil fuel emissions of 0.12 Gt C. For practical purposes, such areas of reforestation would seem to be the maximal ones possible; it will take several decades to get such huge amounts of new forests established. Thus, reforestation may assist in reducing the annual increase in atmospheric carbon content, but it cannot diminish it in absolute terms if no further measures are taken to reduce the man induced CO_2 emissions. However, reforestation is needed for many other important reasons as well, and all contributions to controlling the atmospheric CO_2 concentration are welcome. The CO_2 problem is therefore one more strong reason to stimulate increased rates of reforestation and forest conservation.

REFERENCES

1. Ajtay, G.L., P. Ketner, and P. Duvigneau. (1979). 'Terrestrial primary production and phytomass', in: B. Bolin, E.T. Degens, S. Kempe and P. Ketner (eds.), The Global Carbon Cycle. SCOPE report 13, Wiley, New York, USA, pp. 129-187.
2. Goudriaan, J. and P. Ketner. (1984). 'A simulation study for the global carbon cycle, including man's impact on the biosphere', in: Climatic Change 6, pp. 167-192.
3. Goudriaan, J. (1987). 'The biosphere as a driving force in the global carbon cycle', in: Netherlands Journal Agricultural Science 35, pp. 177-187.
4. Armentano, T.V., and C.W. Ralston. (1980). 'The role of temperate zone forests in the global carbon

cycle', in: <u>Canadian Journal Forest Research</u> 10, pp. 53-60.

5. Brown, S., A.E. Lugo and B. Liegel. (1980). <u>'The role of tropical forests on the world carbon cycle'</u>, Dep. of Energy. DOE-Report No. 007, Washington, USA.

6. Brown, S., A.E. Lugo and J. Chapman. (1985). 'Biomass of tropical tree plantations and its implications for the global carbon budget', in: <u>Canadian Journal Forest Research</u> 16, pp. 390-394.

7. Woodwell, G.M. and R.A. Houghton. (1977). 'Biotic influences on the world carbon budget, in: Stumm, W. (ed.) <u>'Global chemical cycles and their alteration by man'</u>. Report of the Dalem Workshop, November 15-19, 1976, pp. 61-72. Dalem Konferenzen, Berlin.

8. Houghton, R.A., W.H. Schlesinger, S. Brown and J.F. Richards. (1985). 'Carbon dioxide exchange between the atmosphere and terrestrial ecosystems', in: J.R. Trabalka (ed.) <u>'Atmospheric carbon dioxide and the global carbon cycle'</u>. Department of Energy, DOE/ER 0239, Washington DC, pp. 113-140.

9. Seiler, W., and P.J. Crutzen. (1980). 'Estimates of gross and net fluxes of carbon between the biosphere and the atmosphere from biomass burning', in: <u>Climatic Change</u> 2, pp. 207-247.

10. Detwiler, R.P., and C.A.S. Hall. (1988). 'Tropical forests and the global carbon cycle', in: <u>Science</u> 239, pp. 42-47.

11. Bouwman, A.F. (1989). 'The role of soils and land use in the greenhouse effect', in: <u>Netherlands Journal Agricultural Science</u> 37(1), pp. 13-19.

12. Houghton, R.A., R.D. Boone, J.R. Fruci, J.E. Hobbie, J.M. Melillo, C.A. Palm, B.J. Peterson, G.R. Shaver, G.M. Woodwell, B. Moore, D.L. Skole and N. Myers. (1987). 'The flux of carbon from terrestrial ecosystems to the atmosphere in 1980 due to changes in land use: geographic distribution of the global flux', in: <u>Tellus</u> 39B, pp. 122-139.

13. Lugo, A.E., and S. Brown. (1986). 'Steady state terrestrial ecosystems and the global carbon cycle', in: <u>Vegetation</u> 68, pp. 83-90.

14. Esser, G. (1987). 'Sensibility of global carbon pools and fluxes to human and potential climatic impacts', in: <u>Tellus</u> 39B, pp. 245-260.

15. Bolin, B. (1977). 'Charges of land biota and their importance for the carbon cycle', in: <u>Science</u> 196, pp. 613-615.

16. Woodwell, G.M., R.H. Whittaker, W.A. Reiners, G.E. Likens, C.C. Delwiche and D.B. Botkin. (1978). The biota and the world carbon budget', in: <u>Science</u>

199, pp. 141-146.

17. Hampicke, U. (1979). 'Net transfer of carbon between the land biota and the atmosphere, induced by man', in B. Bolin, E.T. Degens, S. Kempe and P. Ketner (eds.) 'The global carbon cycle'. SCOPE report 13, Wiley, New York, USA, pp. 219-236.

18. Houghton, R.A., J.E. Hobbie, J.M. Melillo, B. Moore, B.J. Peterson, G.R. Shaver and G.M. Woodwell. (1983). 'Changes in the carbon content of terrestrial biota and soils between 1860 and 1980: a net release of CO_2 to the atmosphere', in: Ecological Monograph 53, pp. 235-262.

19. Peng, T.H., W.S. Broecker, H.D. Freyer and S. Trumbore. (1983). 'A deconvolution of tree ring based ^{13}C record', in: Journal of Geophysical Research 88(C6), pp. 3609-3620.

20. Emanuel, W.R., G.G. Killough, M. Post and H.H. Shugart. (1984). 'Modelling terrestrial ecosystems in the global carbon cycle with shifts in carbon storage capacity by land-use change', in: Ecology 65(3), pp. 970-983.

21. Peng, T.H. (1986). 'Land-use change and carbon exchange in the tropics: II Estimates for the entire region: comment', in: Environmental Management 10, pp. 573-575.

22. Postel, S., and L. Heise. (1988). 'Reforesting the earth'. Worldwatch Paper No. 83.

23. Hummel, F.C., W. Palz and G. Grassi (eds.). (1988). 'Biomass forestry in Europe: A Strategy for the Future'. Elsevier, London, UK.

24. Dobson, A., A. Dolly and D. Rubenstein. (1989). The greenhouse effect and biological diversity, in: 'Trends in Ecology and Evaluation' 4(3), pp. 64-68.

25. Departments of Forestry Hinkeloord (1984). 'Introduction: towards a global forestation strategy', in K.F. Wiersum (ed.) Strategies and Designs for Afforestation, Reforestation and Tree Planting. PUDOC, Wageningen The Netherlands, pp. 7-25.

26. Bramryd, T. (1982). 'Fluxes and accumulation of organic carbon in urban eco-systems on a global scale', in: R.Bornkamm, J.A. Lee and M.R.D. Seaward (eds.), Urban Ecology. Blackwell, Oxford, UK, pp. 3-12.

27. Wiersum, K.F. (ed.). (1984). 'Strategies and designs for afforestation, reforestation and tree planting'. PUDOC, Wageningen, The Netherlands.

28. FAO (1983). 'Fuelwood supplies in the developing countries'. FAO Forestry Paper No. 42, Rome, Italy.

29. FAO (1982). 'World forest products demand and supply 1990 and 2000'. FAO Forestry Paper No. 29, Rome, Italy.
30. Grainger, A. (1988). 'Estimating areas of degraded tropical lands requiring replenishment of forest cover', in: International Tree Crops Journal 5, pp. 31-61.
31. WRI (1985) 'Tropical forests: a call for action'. World Resources Institute, Washington D.C., USA, 3 vols.
32. Evans, J. (1986). 'Plantation forestry in the tropics - trends and prospects', in: International tree crops Journal 4, pp. 3-15.
33. WRI (1988) 'World resources 1988-89'. A report by the World Resource Institute and the International Institute for Environment and Development. Basic Books, Inc. New York.
34. COFO (1985). 'Tropical forestry action plan'. Committee on forest development in the tropics. FAO, Rome, Italy.
35. FAO (1985). 'Tree growing by rural people'. FAO Forestry Development Paper No. 64, Rome, Italy.

THE RECOVERY OF CARBON DIOXIDE FROM POWER PLANTS

C.A. HENDRIKS, K. BLOK, W.C. TURKENBURG
Department of Science, Technology and Society
University of Utrecht
Oudegracht 320
NL-3511 PL Utrecht
The Netherlands

ABSTRACT. The recovery of carbon dioxide from power plants is technically feasible with techniques currently in use in the chemical industries.

For a pulverized coal plant 90% of the carbon dioxide can be recovered using a chemical absorption process to clean up the flue gases. As a consequence the plant efficiency will decrease from 40.0 to 29.3% and the production costs of electricity will increase from Dct 7.0 to 11.7 per kWh. From these figures a cost of Dfl 64 per ton of carbon dioxide avoided can be calculated. If a gas fired combi-plant is equipped with the same carbon dioxide recovery technique the efficiency falls from 48 to 41.6% while the production costs increase from Dct 6.3 to 8.3 per kWh. The costs are 52 - 59 Dfl per ton of carbon dioxide avoided.

For an integrated coal gasification combined cycle plant (IGCC) the cost-effectiveness of carbon dioxide recovery can be much better. In this case a physical absorption process is used to clean the synthesis gas. The efficiency will probably decrease from 43.6 to 38.2%. The electricity price will increase from Dct 7.0 to 8.7 per kWh. The costs are Dfl 26 per ton carbon dioxide avoided.

Other ways to recover carbon dioxide as well as possibilities to improve current techniques have to be investigated, which might result in more cost-effective options.

1. INTRODUCTION

The emission of carbon dioxide can be reduced in many ways. Reduction can take place for example by energy savings or by a larger input of non-carbon-dioxide-emitting energy sources like renewable energy and nuclear energy. Emitted carbon dioxide can also be compensated by afforestation (or reforestation) of parts of the earth.

Recovery and disposal of carbon dioxide from combustion processes is another opportunity to reduce the emission of carbon dioxide into the atmosphere. In this article we will describe several methods to recover carbon dioxide. Furthermore we will make estimates of the costs associated with the recovery of carbon dioxide. After recovery the carbon dioxide has to be disposed of. There are several ways of carbon dioxide disposal of which the storage in exhausted natural gas reservoirs seems most promising at the moment [1,2,3]. We will not deal with this topic here and refer to the article of Van der Harst en Nieuwland elsewhere in this publication.

We consider two processes of carbon dioxide recovery in detail:
- The recovery of carbon dioxide from the flue gases of a conventional gas or coal fired power plant using a chemical absorption process according to the configuration proposed by Steinberg [4].
- A newly proposed configuration of carbon dioxide recovery from the syngas of an integrated gasification combined cycle plant (IGCC) using a shift reaction and a physical absorption process [1].

Furthermore, we will describe a number of other options for the recovery of carbon dioxide. Finally, we will draw conclusions on the role carbon dioxide recovery could play in the reduction of the greenhouse effect.

Carbon dioxide recovery can be applied to all kinds of combustion processes. In general however, due to scale-effects, the recovery will be economically most attractive at locations with a high carbon dioxide production density. Also other conditions can be important, e.g., the presence of low-grade waste heat. In this article we will concentrate on the recovery of carbon dioxide from fossil-fuel fired power plants.

All financial figures in this article are presented in Dutch guilders (Dfl 1 = $ 0.45 = ECU 0.43). The capital costs are calculated using depreciation over the lifetime of the installations (generally 25 years) and a real interest rate of 4%. When US cost data are used a 30% discount is applied to compensate for the lower costs in Europe of the type of installations described here.

In the calculations current fuel prices are used, i.e. Dfl 4 per GJ for coal and Dfl 6 per GJ for natural gas. To analyse the sensitivity calculations have also been performed for a doubling of these fuel prices.

2. THE RECOVERY OF CARBON DIOXIDE FROM THE FLUE GASES OF A
CONVENTIONAL GAS OR COAL FIRED POWER PLANT USING A CHE-
MICAL ABSORPTION PROCESS

The concentration of carbon dioxide in the exhaust
gases of conventional coal, oil or gas fired power plants
is low: 8 - 15% by volume. For such concentrations chemical
absorption processes seem most suitable up to now. In a
chemical absorption process the carbon dioxide is recovered
by leading the flue gas through a solution containing the
absorber.
The recovery process which we will describe here consists
of three steps:
- absorption of carbon dioxide by a MEA solution;
- recovery of the carbon dioxide from the MEA solution by
  using low grade heat extracted from the power plant;
- compression and drying of the carbon dioxide.
A process like this has been described by Steinberg et al.
[4].

2.1 Absorption of carbon dioxide by a MEA solution

Of the chemical absorbers monoethanolamine (MEA) is
the one which is applied mostly. The MEA absorption/desorp-
tion process can be described as follows:

$$R-NH_2 + H_2O + CO_2 \underset{\longleftarrow \text{ } 150 \text{ °C}}{\overset{27 \text{ °C} \longrightarrow}{\longleftrightarrow}} R-NH_3 HCO_3$$

The presence of oxygen (causing oxidation of the MEA) and
vapour (carbon dioxide and vapour form a corrosive mixture)
in the flue gases makes improvements in the MEA process
necessary. Dow Chemical has developed the so-called MEA-FT
process, using as its base the MEA absorber. By using in-
hibitors, the presence of oxygen can be tolerated up to 8%
by volume. The corrosion problems are also reduced, resulting in higher intake capacity. The resulting absorber solution is called Dow Gas/Spec FS-1. Together with SO_2, MEA
forms an insoluble non-regenerable salt. At power plants
where SO_2 is released, flue gases should be desulphurated
before entering the carbon dioxide recovery plant.
A flow diagram of the chemical absorption process is
shown in figure 1. Flue gases enter the recovery system at
a pressure slightly above the ambient pressure and a temperature generally in excess of 120 °C. First, the gas is
compressed to somewhat less than 1.3 bar to accommodate the
pressure drop through the system, and the gas is cooled to
nearly 50 °C. Next the gas containing carbon dioxide flows
through the MEA solution. The carbon dioxide binds chemically to the absorber. The solution containing the absorbed

carbon dioxide then flows to another drum, where it is heated. In this drum the carbon dioxide desorbs.

A practical efficiency for the recovery of the carbon dioxide from the flue gases is 90%. The recovery of carbon dioxide absorbed by the MEA solution can be fully recovered. So the overall carbon dioxide recovery efficiency is 90%.

Investment and maintenance costs for the MEA-units are derived from [4] and given in tabel 1.

The costs of supplying fresh MEA solution amount to Dfl 3000 per ton (causing a cost of Dfl 6 per ton carbon dioxide recovered). Used MEA can be disposed of by combustion in refuse incinerators. The combustion follows the standard procedures for Dutch refuse incinerators. This way of MEA disposal costs around Dfl 200 per ton (Dfl 0.40 per ton carbon dioxide recovered). The disposal costs turn out to form only a small proportion of the total costs.

2.2 Recovery of CO_2 by using extracted heat

Heat forms the largest part of the energy input for the desorption process. According to the literature the heat, which is necessary for the desorption of carbon dioxide by MEA, equals 4.2 - 5.3 GJ per ton recovered carbon dioxide [4,5,6]. The heat input depends on the flow of the MEA solution. This flow depends on the intake capacity of the solution and the concentration of carbon dioxide in the flue gas. The heat can be generated in separate steam boilers, but a far more energy-efficient and cost-effective method is extraction of the heat in the form of low pressure steam from the power plant. Figure 2 shows a diagram for integrated carbon dioxide recovery in the power plant itself. As a consequence of this extraction, there will be less steam available for electricity production. This will cause the overall efficiency of the power plant to diminish.

The power loss can be calculated as follows. For each ton of carbon dioxide recovered we assume a heat requirement of 5 GJ. This heat can be obtained by extracting 2.06 tons steam of 2 bar/180 °C from the steam cycle of the power plant (enthalpy of 2.84 MJ/kg) and cooling it to water of 90 °C (enthalpy 0.41 MJ/kg). The same amount of steam could have been used to generate 260 kWh of electricity. This amount of electricity not produced leads to a reduction of approx. 20% for coal and 10% for gas in the power production of the turbine/generator.

In this study we started from a power plant configuration that exists nowadays. It should be noted that the integration of the absorption/desorption cycle in a power plant can possibly be optimized. It is, for example, not necessary to use superheated steam for the extraction. Furthermore the 90 °C return water from the desorption plant might be utilized in a well designed plant.

2.3 Compression and drying of CO_2

We assume that compression to 60 bar is necessary before the carbon dioxide can be transported or otherwise utilized.
The compression process is done most effectively by alternatively compressing and cooling of the carbon dioxide flow. For such a three-stage compression process we developed the following relation for the energy consumption:

$$P(kWh_e/ton\ CO_2) = 18.8\ ln\ (p_{high}/p_{low})$$

in which P is the electricity required while p_{high} and p_{low} are the end and begin pressure respectively. The energy needed for the compression (compressibily rate of 3.9) to 60 bar amounts to 77 kWh_e per ton CO_2.
For the coal-fired plant considered here the compressor could consist of two trains in parallel. In that case the costs will amount to Dfl 120 million. For the natural gas fired plant under consideration half of these costs are sufficient. The cost data given here include motor, control, infrastructure, installation etc. They were provided by a manufacturer [7].
Drying of the carbon dioxide is required because the combination of water and carbon dioxide makes up a corrosive solution and the formation of hydrates should be prevented. Most of the water is removed during the compression. After each compression stage the condensate is removed in knock-out-drums. After the last stage the remaining water is removed by the solvent tri-ethylene-glycol. Regeneration of this solvent takes place by air stripping.
At last some electricity is needed for pumping, control etc. These additional energy requirements amount to approx. 13 kWh_e per ton CO_2.

2.4 Overview

Table 1 gives a summary of the effect of carbon dioxide recovery on the power plants energy balance and the production costs. We consider two types of power plants:
1. A conventional coal-fired steam boiler with an efficiency of 40% without carbon dioxide recovery.
2. A natural gas fired power plant. Nowadays newly built plants will generally be a combined cycle plant without additional firing in the waste heat boiler. The efficiency of such a plant can be approx. 50%. Such a plant, however, has a very low carbon dioxide content (approx. 3%) in its flue gases, which makes it less cost-effec-

tive to recover carbon dioxide.

Therefore, we derive the gas fired plant with CO_2 recovery from a so-called combi-plant: a gas-fired power plant consisting of a gas turbine feeding its exhaust gases to a fully fired boiler. The efficiency of such a plant is generally somewhat lower [8]: 48%.

In both cases we start with a 600 MW_e conventional power plant. From this plant we derive a plant with carbon dioxide recovery which removes 90% of the initial carbon dioxide. In doing this, we keep the fuel input the same.

From our calculations we conclude that the efficiency of the coal- and gas-fired plant fall with respectively 10.7 and 6.4 %-points. The overall production costs rises from Dct 7.0 to 11.7 per kWh for the coal-fired plant and from Dct 6.3 to 8.3 per kWh for the gas-fired plant. The additional costs per ton carbon dioxide avoided are Dfl 64 for the coal-fired plant and Dfl 52 for the gas-fired plant. It should be noted that the amount of carbon dioxide avoided is not the same as the amount which is recovered, because part of it would not have been produced if no carbon dioxide recovery were applied.

In this comparison we compared the gas fired combi plant with carbon dioxide recovery with a combi plant without recovery (efficiency 48%). If, however, we compare the combi plant with recovery with a combined cycle plant without recovery (efficiency 50%) the cost penalty is somewhat higher: an electricity price increase of Dct 2.1 per kWh and a cost of Dfl 59 per ton of carbon dioxide avoided.

The same calculations can be made with doubled fuel prices, so Dfl 8 per GJ for coal and Dfl 12 per GJ for gas. In that case the calculated kWh_e prices for coal- and gas-fired power plants are resp. Dct 10.6 and 10.5 without carbon dioxide recovery and Dct 16.6 and 13.4 with carbon dioxide recovery. The costs for carbon dioxide recovery are Dfl 82 for coal and Dfl 70 - 83 for gas per ton carbon dioxide avoided.

Table 1. The power balance and cost calculations of carbon dioxide recovery from a conventional coal- and a gas-fired power plant. All cost figures are given in Dfl.

| | COAL PLANT | | GAS PLANT | |
	without recovery	with recovery	without recovery	with recovery
Fuel input (MW_t)	1,500	1,500	1,250	1,250
Power loss by steam extraction (MW_e)		-119		-59
Gross capacity (MW_e)	600	481	600	541
Power demand pumping etc. (MW_e)		-6		-3
Power demand for compression (MW_e)		-35		-17
Net capacity (MW_e)	600	440	600	520
Overall efficiency (LHV, %)	40	29.3	48	41.6
Recovered carbon dioxide (kton/yr)		2,740		1,360
Plant investment costs (million)	1,200	1,200	750	750
MEA-absorption unit (million)		280		161
Compressor (million)		120		60
Total investment costs (million)	1,200	1,600	750	970
Specific investment costs (Dfl/kW_{net})	2,000	3,640	1,250	1,870
Plant O&M costs (million/yr)	44	44	18	18
Additional O&M costs (million/yr)		13		7
MEA costs (million/yr)		18		9
Total O&M costs (million/yr)	44	76	18	34
Specific O&M costs (Dfl/kW_n et.year)	74	172	30	64
Yearly exploitation at 6000 running hours				
Capital costs (million/yr)	77	102	48	62
O&M costs (million/yr)	44	76	18	34
Fuel costs (million/yr)	130	130	162	162
Electricity price (Dct/kWh)	7.0	11.7	6.3	8.3
Costs of carbon dioxide recovery (Dct/kWh)		4.9		2.0
Avoided carbon dioxide (kton/yr)		1,930		1,160
Cost effectiveness (Dfl/ton CO_2 avoided)		64		52

3. THE RECOVERY OF CARBON DIOXIDE FROM THE SYNGAS OF AN IN-TEGRATED GASIFICATION COMBINED CYCLE PLANT (IGCC) USING A SHIFT REACTION AND A PHYSICAL ABSORPTION PROCESS

A promising option for electricity generation is the integrated use of a gasifier and combined cycle plant (IGCC). A coal gasifier converts coal to a synthesis gas, of which hydrogen and carbon monoxide are the main components.
The recovery technique of the carbon from this syngas is, because of the high carbon concentration, much easier than the recovery of carbon from the exhaust gases. Recently we proposed a new configuration for carbon dioxide recovery which takes place in three steps [1]:
a. conversion of the carbon monoxide to carbon dioxide by using the monoxide shift reaction;
b. recovery of carbon dioxide from the syngas by using a physical absorption process with selexol as absorber;
c. compression and drying of the carbon dioxide.
The technology for the removal of carbon dioxide from gasifiers corresponds to techniques that are used, for example, in the ammonia industry. Thus, the implementation of carbon dioxide recovery facilities can be done without major technological problems.

3.1 Base configuration

The original design is based on an integrated gasification combined cycle (IGCC) plant using the Shell gasification process. The basic configuration of such a plant is described in [9]. The principle of coal gasification lies in the chemical conversion of the energy of the coal into a more suitable form: CO, H_2 and heat. The heat is used to produce high-pressure steam for expansion in a steam turbine and the CO/H_2 mixture is combusted with air in a gas turbine. The exhaust gases of this gas turbine are used to produce steam in a waste heat boiler. This steam is also fed to the steam turbine.
To recover the carbon from the syngas, this configuration is extended with a shift reactor and a carbon dioxide removal unit. These units are placed in the syngas stream after the acid gas removal and before the gas turbine. This is necessary as the catalysts of the shift reactor are very sensitive to sulphur. The acid gas removal unit removes about 95-99% of the original amount of sulphur. The final fraction of the remaining sulphur will be removed from the syngas together with the carbon dioxide. The absorber selexol has a greater affinity for H_2S than for carbon dioxide. Thus almost 100% desulphurization is the final result.

3.2 The shift reactor

During the shift reaction CO and steam convert to CO_2 and H_2. Thereby the chemical energy is transferred to H_2 and heat. The conversion can be presented as follows:

$$H_2O(l) \longleftrightarrow H_2O(g)$$

$$CO(g) + H_2O(g) \longleftrightarrow CO_2(g) + H_2(g)$$

The positive enthalpy of the shift reaction, dependent on the temperature, is the same but opposite in sign as the vaporization enthalpy of water. The overall heat effect for the total shift reaction is roughly zero. The net result is that, after the conversion, the energy content (lower heating value) of the syngas is decreased by about 8.5%. In theory this energy loss could be regained by condensing the water vapour in the exhaust of the combined cycle plant. In practice, however, it is very difficult to do this in a power plant, so we consider this energy as lost.

The carbon dioxide removal process is depicted in figure 3. The conversion rate of carbon monoxide to carbon dioxide does not depend on the pressure. The original gasification takes place at 24 bara. It appears reasonable to execute the conversion at the same pressure. The shift reaction takes place in two stages.

After the acid gas removal the syngas is at a temperature of 40 °C. This gas will be reheated to 350 °C, where it passes a saturator. After the saturator the gas stream is heated to 400 °C, the operational temperature of the HT-shift reactor. A small difference in temperature is necessary to prevent condensation of water in the shift reactor, which can cause poisoning of the catalyst. The catalyst used in this reactor is based on an iron and chrome alloy. If stoichiometric rates of H_2O and CO are present, 78% of the CO is converted to CO_2 in the high-temperature conversion.

The second shift reactor operates at a temperature of 220 °C (LT-conversion). The low temperature catalyst contains copper and zinc. This alloy is very sensitive to sulphur. In this reactor another 12% of the original carbon monoxide is converted. Therefore the total carbon monoxide conversion efficiency amounts to more than 90% [10,11].

In our calculations we depart from a power plant with a syngas stream of 415 Nm^3/h (dry). The shift reactor requires an investment of Dfl 50 million [12]. The catalyst volume required is 230 m^3 for each stage. The purchase costs of the catalysts are aprox. Dfl 18,000/m^3 for the LT-stage and Dfl 17,000/m^3 for the HT-stage [13], giving a total catalysts purchase cost of Dfl 8 million. The life of both catalysts can be estimated to be 4 to 6 years. The catalyst costs are counted as O&M costs.

The two shift reactors can operate independently from each other. This means that a good reliablity of the carbon dioxide recovery can be obtained without the need of installing a spare shift train.

A special point of concern is the heat balance of the shift reaction. Although the overall reaction requires no energy, it should be noted that the syngas stream has to be heated first, while afterwards the heat can be recovered again.
The latter can be done by generating high-pressure steam from the heat which is formed at the HT-conversion. The heat released in the LT-conversion can be used to create the medium-pressure steam. About 30% of the conversion heat is needed to heat the syngas up to 400 °C. The recovery of this heat is estimated to be 90%. The total loss would then be about 3%. In further calculations this amount will be neglected. There is also a possibility of using low temperature heat to reheat the syngas and of using the heat released by the conversion for the production of electricity, which might even have a positive effect on the total heat balance of the power plant.

3.3 Carbon dioxide recovery using selexol

The shifted syngas stream contains a relatively high concentration of carbon dioxide: approx. 40% at a high pressure. For such conditions a physical absorber can be used instead of a chemical absorber. In our calculations we will assume that the absorber in use will be selexol, although other absorbers (e.g., sulfinol) might be used as well. Selexol is a 95% solution in water of dimethyl ether of polyethylene glycol. The absorption/desorption process works out as follows.

After the shift reaction the gas-mixture containing carbon dioxide, hydrogen and a small amount of carbon monoxide flows counter to the current of the selexol (unit no 1 in figure 4) [14]. The carbon dioxide is absorbed by the solvent and collected at the bottom of the unit. The solubility of carbon dioxide increases under higher pressures and at lower temperatures. Under such conditions the solvent circulation rate can be low.

The rich solvent flows into the recycle flash drum (3) where essentially all the co-absorbed hydrogen and carbon monoxide are flashed. The carbon dioxide is recovered by reducing the pressure, which takes place in several serial-connected drums (4) en (5). The carbon dioxide is released partly under high pressure and partly under atmospheric pressure. After the desorption stages the selexol contains 25-35% of the original dissolved carbon dioxide. This carbon dioxide is recycled back to the absorber and is recovered in a later cycle.

The carbon dioxide recovery rate from the syngas stream will be approx. 98%, in which figure all losses have been taken into account. As 90% of the carbon was converted to carbon dioxide, this means an overall recovery rate of 88%.

In the selexol recovery process energy is required for the pumping of selexol. The electricity requirements are approx. 17 MW, assuming a pumping efficiency of 70%. About 50% of this energy can be recovered by applying power recovery at the pressure release of the selexol, leaving a net power consumption of 9 MW.

An alternative set-up for the selexol process is to achieve a higher direct recovery rate of carbon dioxide (up to 97%) by flashing the solvent further to vacuum pressures. Whether a vacuum flash drum should be chosen depends strictly on economic considerations: the pumping energy diminishes but the compression energy for carbon dioxide increases.

The costs of the absorption/desorption unit are derived from [14] in which a smaller unit is described. Using a scale-factor of 0.8 we can estimate the investment costs of the reactor described here to be Dfl 90 million. This figure was confirmed by manufacturers data [15].

3.4 Compression and drying of CO2

The desorbed carbon dioxide is released not only at atmospheric pressures, but partly also at elevated pressures. The latter is favourable because of the lower energy requirement for compression. Simplifying we can say that one half of the carbon dioxide is released at 8 bar and the other half at 1 bar. From these figures we can calculate an average electricity demand of the compressor of 53 kWh_e per ton CO_2.

Further remarks concerning the compression and drying of carbon dioxide have already been made in section 2.3. The compressor investment costs are taken Dfl 92 million [7].

3.5 Overview

Table 2 gives a summary of the effect of carbon dioxide recovery on the energy balance and on the production costs of electricity from the IGCC power plant.

The original design, described in [9], has a capacity of 710 MW_e. From this design we derive a power plant from which 88% of the carbon dioxide is recovered but in which the fuel input is kept the same. We conclude that the power falls to 620 MW_e, a decrease of 13%. The efficiency falls from 43.6% to 38.1%, a decrease of 5.5 %-points. The overall production costs rise from Dct 7.0 to 8.7 per kWh. The costs of carbon dioxide recovery are calculated to be

Dfl 26 per ton.

The calculation has also been made with a doubled coal price: Dfl 8 per GJ. The calculated kWh_e prices then increases from Dct 10.3 without carbon dioxide recovery to Dct 12.5 with carbon dioxide recovery. The costs for carbon dioxide recovery then become Dfl 33 per ton carbon dioxide avoided.

TABLE 2. Energy balance and cost calculation of the carbon dioxide recovery from an Integrated Gasification Combined Cycle Plant (IGCC)

	without recovery	with recovery
Fuel input (MW_t)	1,630	1,630
Syngas production (MW_t)	1,316	1,206
Syngas cooler steam production (MW_t)	302	302
HRSG steam production (MW_t)	352	321
Gross power	802	747
Gas turbine power (MW_e)	480	440
Steam turbine power (MW_e)	322	307
Internal power demand (MW_e)	-92	-92
Power demand for selexol pumping (MW_e)		-9
Power demand for compression (MW_e)		-24
Net capacity (MW_e)	710	620
Overall efficiency (LHV, %)	43.6	38.1
Recovered carbon dioxide (kton/year)		2,910
Plant investment costs (million)	1,560	1,560
Shift reactor, excl. catalysts (million)		50
Selexol absorption unit (million)		90
Compressor (million)		92
Total plant investment cost (million)	1,560	1,790
Plant investment costs (Dfl/kW_n et)	2,200	2,890
Plant O&M costs (million/yr)	57	57
Additional O&M and chemicals (million/yr)		11
Catalysts (million/yr)		2
Total plant O&M costs (million/yr)	57	70
Specific plant O&M costs (Dfl/kW_n et.year)	80	112
Yearly exploitation at 6000 running hours		
Capital costs (million/yr)	100	115
O&M costs (million/yr)	57	70
Fuel costs (million/yr)	141	141
Electricity price (Dct/kWh)	7.0	8.7
Costs of carbon dioxide recovery (Dct/kWh)		1.7
Avoided carbon dioxide (kton/year)		2,490
Cost effectiveness (Dfl/ton CO_2 avoided)		26

4. OTHER WAYS OF CARBON DIOXIDE RECOVERY FROM POWER PLANTS

4.1 Using other CO_2 recovery techniques

An other technique to recover carbon dioxide is cryogenic distillation. This technique is most effective at higher concentrations of carbon dioxide (>20%). A process suitable for the recovery of carbon dioxide from combustion gases has been developed by Koch Process Industries: the Ryan/Holmes process [16].
A study performed by Schlüssel and Kümmel [17] shows that removing carbon dioxide from a conventional power plant with cryogenic distillation is less cost-effective than removing carbon dioxide with the chemical absorption process described in section 2.

An option with a future perspective is the use of membranes. These membranes can be used to separate for instance carbon dioxide from combustion gases, carbon dioxide from syngas or oxygen from air. The latter option might be important for the technique described in section 4.2. The big advantage of membrane techniques is their low energy use. A disadvantage of membrane separation techniques is the relative impureness of the separation products. This disadvantage can be avoided by using multi-stage membrane installations or by combination with other separation techniques. Currently large-scale membrane separation is only used for gas flows with a high partial carbon dioxide pressure. Second generation membranes are suggested to have the possibility to be used for more diluted carbon dioxide flows.

4.2 Using oxygen instead of air

A quite different approach is the combustion of coal under oxygen instead of air. The result of this combustion process would be an exhaust gas that consists maninly of (more or less contaminated) carbon dioxide.

It is impossible to feed pure oxygen to the combustion chamber, because of the high flame temperature which would result from this type of combustion. This problem can be solved by partly recirculating the resulting carbon dioxide to the inlet of the combustion chamber and mixing it up with oxygen. The combustion products are carbon dioxide, water, sulphur oxide and nitrous oxides. The water can be removed and the carbon dioxide can be compressed. It might be an advantage that the other contaminants can be removed together with the carbon dioxide. In this process quite a great deal of energy is required to separate the oxygen from the air.

Experimental investigations have been performed by the Argonne National Laboratory [18]. This research was directed especially to retrofit situations, which means that the heat transfer should be approximately the same compared with combustion under air. From the experiments it was concluded that a gas with a CO_2/O_2 ratio of 2.23 till 2.42 turned out to have the desired property.

Design studies has been performed to investigate the feasibility of retrofitting a Scottisch power plant for producing carbon dioxide for enhanced oil recovery at the North Sea [19]. These studies show that the cost effectiveness of this type of carbon dioxide recovery is of the same order of magnitude as the recovery using a chemical absorption process described in section 2.

However, if a new power plant is built that has been designed such that carbon dioxide is recovered the costs might well turn out to be lower. An option is the pressurized combustion which might avoid compression costs. However, a combustion chamber that fits has to be developed yet.

5. CONCLUSIONS

From our calculations we conclude that carbon dioxide emissions from conventional power plants can be avoided at a cost of Dfl 64 per ton for a coal fired power plant and Dfl 52 - 59 per ton for a gas fired power plant. As we have shown carbon dioxide recovery from an integrated gasification combined cycle plant (IGCC) can be much cheaper: Dfl 26 per ton.

At current environmental demands the electricity production costs for a pulverized coal power plant seem to be somewhat lower than for an IGCC. If we take carbon dioxide recovery into account this will surely no longer be the case. Although the electricity price increase for CO_2 from an IGCC of approx. Dct 2 per kWh is substantial, it equals the order of magnitude of the price increase due to the removal of the acidifying components SO_2 and NO_x.

The costs of carbon dioxide recovery can probably be decreased if new technologies can be applied, like the use of advanced membranes and the pressurized combustion under a mixture of O_2 and CO_2.

Compared with other possible options for the reduction of carbon dioxide emissions it can be concluded that, although a broad range of energy conservation options surely will turn out to be cheaper, the costs of carbon dioxide recovery, including the costs of disposal [1,2], might in lots of cases be competitive to the additional costs if other carbon dioxide reduction technologies are applied like some renewable options, nuclear energy and, depending on the fuel prices, shifting from coal to natural gas [20,21].

ACKNOWLEDGEMENTS

We would like to thank the Netherlands Energy Research Foundation and the Dutch Ministry of Housing, Physical Planning and the Environment for the financial support of the work reported here.

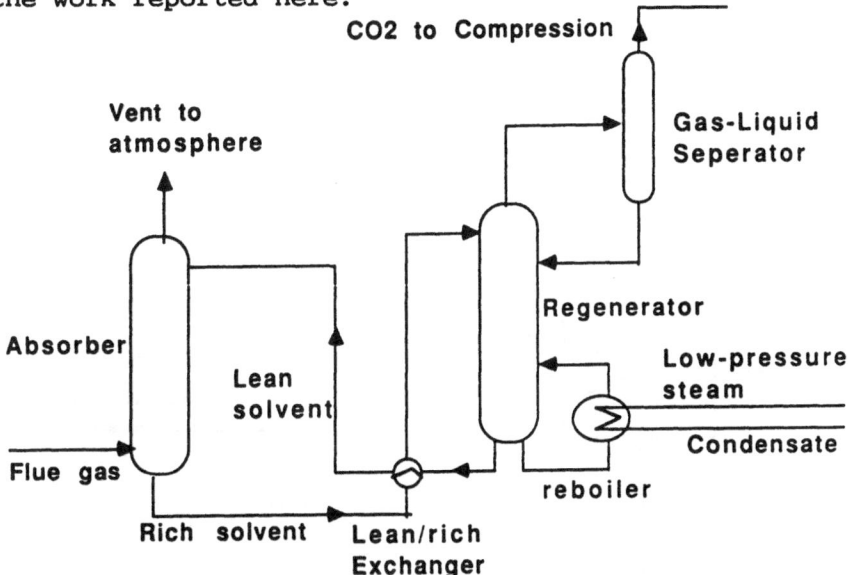

Figure 1. Diagram of the chemical absorption/desorption cycle using a MEA solution.

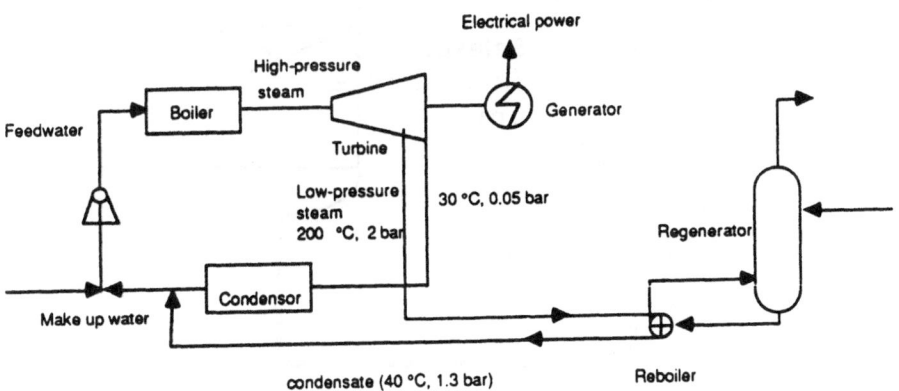

Figure 2. The integration of the chemical absorption process into the power plant.

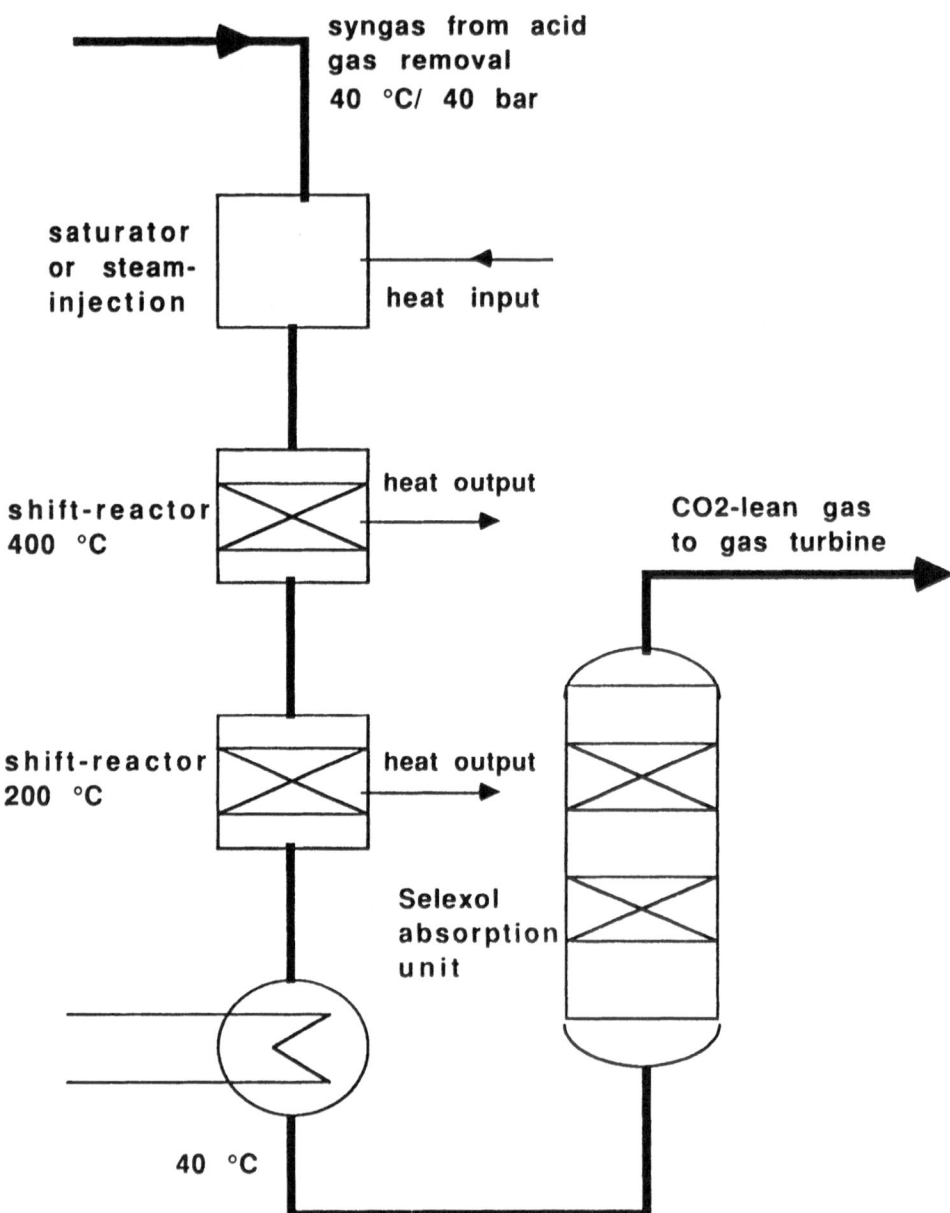

Figure 3. Diagram of the process which recovers carbon dioxide from the syngas.

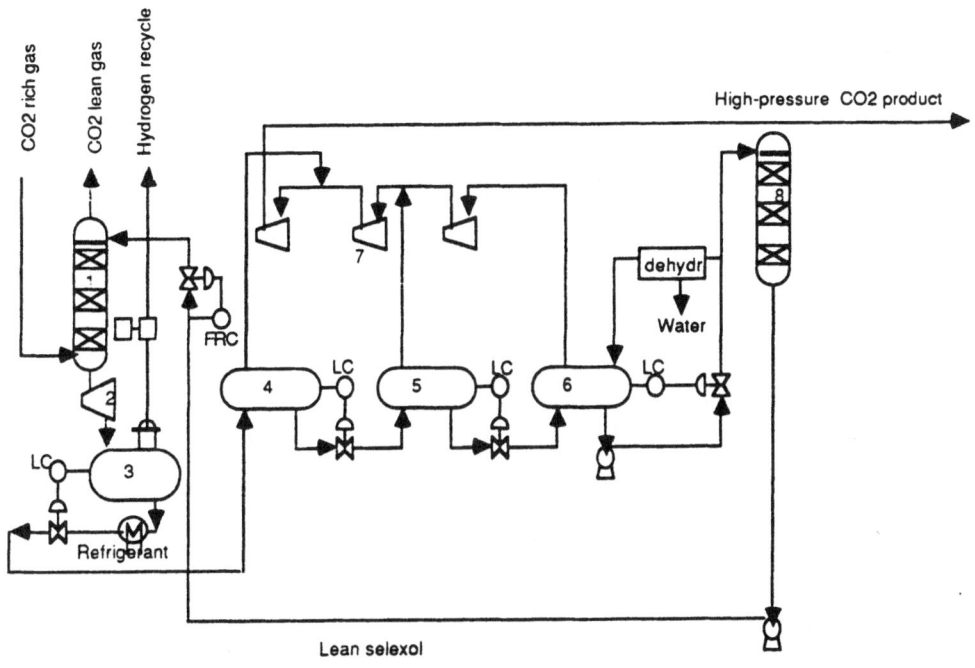

Figure 4. Diagram of the physical absorption process using a selexol solvent.

REFERENCES

1. K. Blok, C.A. Hendriks, W.C. Turkenburg: The Role of Carbon Dioxide Removal in the Reduction of the Greenhouse Effect, IEA/OECD Expert Seminar on Energy Technologies for Reducing Emissions of Greenhouse Gases, Paris, 12-14th April 1989.
2. A.C. van der Harst, A.J.F.M. van Nieuwland: Disposal of Carbon Dioxide in Depleted Natural Gas Reservoirs (elsewhere in this publication).
3. E.H. Lysen: The Absorption of Carbon Dioxide by the Oceans, Lysen Raadgevend Ingenieursbureau, Amersfoort, February 1989 (in Dutch).
4. M. Steinberg, H.C. Cheng: A System Study for the Removal, Recovery and Disposal of Carbon Dioxide from Fossil Fuel Power Plants in the US, Brookhaven National

Laboratory, February 1985.

5. C.R. Pauley: CO_2 Recovery from Flue Gas, CEP, May 1984.

6. J.J. Taber: Need, Potential and Status of CO2 for Enhanced Oil Recovery, from: Recovery of Carbon Dioxide from Man-Made Sources, Proceedings of a Workshop held in Pacific Grove, California, 11-13 February 1985, pp. 11-32.

7. Personal communication mr. De Vries, Sulzer Nederland BV. The cost data provided are approximately half the compressor costs used by Steinberg [1].

8. A. Steiner: Combined Gas-Steam Turbine Power Plants, Presentation at the ASME Cogen-Turbo, Montreux, September 2-4, 1987.

9. Integrated Coal Gasification Power Plant Study, Fluor Daniel, Haarlem, February 1988.

10. J.M. Blanken: The Ammonia Industry (1), Procestechnologie nr 2, 1988, pp. 37-42 (in Dutch).

11. Riegel's Handbook of Industrial Chemistry, J.A. Kent (ed.), 7th ed, Van Nostrand Reinhold Company, New York, 1974, pp. 78-95.

12. Personal communication mr. Meent, Comprimo Engineers and Contractors, Amsterdam.

13. Personal communication mr. Vliegen, DSM, Sittard, The Netherlands.

14. V.A. Shah, J. McFarland: Low Cost Ammonia and CO_2 Recovery, Process Technology, March 1988, pp. 43-46.

15. Written communication mr. J.G.H. van Leeuwen, Norton CPP, Zoetermeer.

16. A.S. Holmes: Process Improved Acid Gas Separation, Hydrocarbon Processing, May 1982, p. 131.

17. U. Schlüssel, R. Kümmel: CO_2 Removal from Flue Gases of Fossil Fuel Power Plants by Refrigeration under Pressure, Physkalischer Institüt, Universität Würzburg, Germany.

18. A.M. Wolsky: A New Approach to CO_2 Recovery from Combustion, Proceedings of a workshop held in Pacific Groove, California, February 11-13, 1985, pp. 76-81.

19. Offshore Business 1985/6, Vol. 2: Enhanced Oil Recovery, Smith Rea Energy Associated Ltd., London, 1986.

20. P.A. Okken, T. Kram: Two 'Low CO_2' Scenario's for The Netherlands, IEA/OECD Expert Seminar on Energy Technologies for Reducing Emissions of Greenhouse Gases, Paris, 12-14th April 1989.

21. K. Blok: Data acquisition of energy conservation techniques for the year 2010, Dept. of Science, Technology and Society, University of Utrecht, June 1989 (in Dutch).
K. Blok: A Dutch Supply Curve for Energy Conservation Techniques, ETSAP-workshop of the International Energy Agency, Paris, June 22-28, 1989.

STORAGE OF CARBON DIOXIDE IN THE OCEANS

Hein J.W. de Baar and Michel H.C. Stoll
Climate Research Unit
Netherlands Institute of Ocean Sciences (NIOZ)
P.O. Box 59
1790 AB Den Burg, Texel
The Netherlands

ABSTRACT.   Past and present variations of the global C cycle are dominated by the oceans. At present the oceans act as moderator for the atmospheric CO_2 increase from combustion of fossil fuels. From the annual fossil fuel CO_2 emission by mankind of 5-6 GtC/yr about half or 2-3 GtC/yr is currently taken up by the oceans. The rate of CO_2 transfer through the surface ocean, which acts as a barrier between atmosphere and deep ocean, is critical. The transfer routes are through both inorganic dissolution and biological fixation. We are currently not certain which route is dominant. Research towards worldwide quantification of both routes is now underway in the Joint Global Ocean Flux Study. The strictly inorganic route will definitely decrease in the future, the biological route may well remain more or less constant or increase in the future. More accurate quantification of both routes is crucial for improving accuracy of predictive CO_2 climate models upon which policy decisions for curtailing CO_2 emissions are to be based.

The crucial problem with fossil fuel CO_2 is its very rapid introduction within 100-200 years into the atmosphere as opposed to the very slow response of many thousands to millions of years of the deep ocean in absorbing such CO_2. Eventually the capacity for storage of CO_2 in the deep ocean is very large. Yet in the meantime we will witness a transient peak of atmospheric CO_2 which may yield catastrophic changes in the climate. Only after several thousands to millions of years most, but not all, of the fossil fuel CO_2 will be taken up by the oceans.

Deep sea injection has been proposed as a technical fix for bypassing the surface ocean barrier in order to delay the peak buildup of atmospheric CO_2. Only CO_2 from large stationary energy plants (currently 30% of total emission) is suitable for deep sea injection. At the expense of 30-45% of the total energy produced this results in a 30-45% decrease in electricity production. Furthermore ocean circulation would also return the deep injected CO_2 to the surface within decades or centuries. Deep sea injection is at best a partial, expensive and temporal remedy to the CO_2 problem. With or without deep sea injection the intrusion of CO_2 into first the surface waters and then the deep ocean will cause shifts in ecological conditions. For example the acidity of surface waters would roughly double, causing inevitable but poorly predictable shifts in the plankton population. Similar shifts in the deep sea ecosystem, also due to dissolution of calcite sediments, would be accelerated by deep sea CO_2 injection.

Reduction of CO_2 production by both energy conservation as well as shifting to other energy sources is recommended instead.

1. THE ROLE OF THE OCEANS IN THE GLOBAL CARBON CYCLE

1.1 The global ocean

The ocean covers most of the surface of the globe. Its metabolism, i.e. the internal cycling and boundary exchanges of chemical constituents of the ocean, has a major impact on the atmosphere and the terrestrial biosphere. One example is the cycle of Carbon, the prime building block of all biomass (Figure 1). The seawater reservoir contains about 40,000 x 10^{15} grams or 40,000 Gigaton Carbon (largely as the bicarbonate ion: HCO_3^-) and its annual exchange rate with the atmosphere is of the order of 100 GtC per year. The atmospheric reservoir itself contains only about 700 GtC as carbon dioxide (CO_2) and exchanges at a rate of about 70 GtC/yr with the terrestrial biosphere. The latter contains about 2,000 GtC, of which some 70 per cent, approximately 1,400 GtC in the soil (humus) with the remaining 600 GtC above ground as plants.

Throughout geological time there have been variations in the geosphere/biosphere system, including shifts in the above inventories and fluxes of the C system. For example, records over the past 160,000 years derived from Antarctic ice cores (Figure 2) show atmospheric CO_2 variations between about 120 ppm and 270 ppm, the latter correlating with variations in surface temperature. The apparent cause/effect relationship of this glacial/interglacial CO_2 shift is the subject of considerable debate (1-8). Yet there is no doubt that the oceans, through variations in circulation, biological productivity, nutrient regime, and a number of other factors, play a crucial role (9-14).

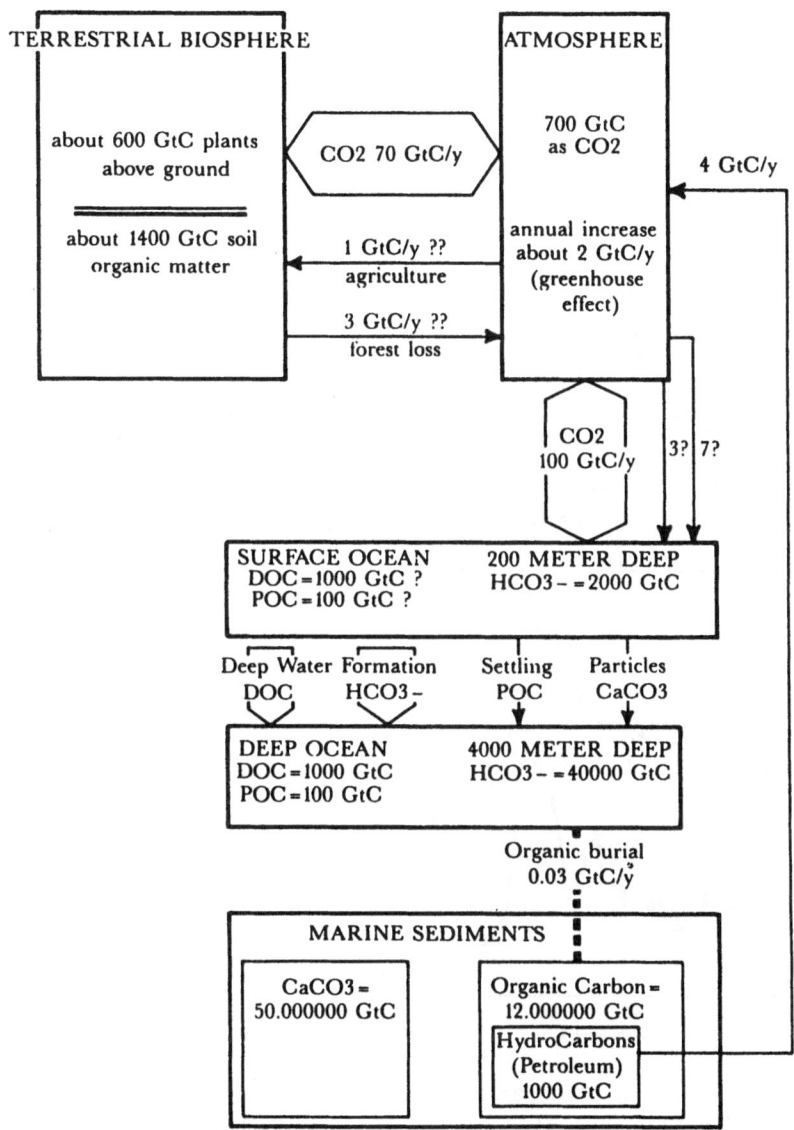

Figure 1. The global Carbon budget (after ref. 10). Inventories [1 Gigaton Carbon = 1,015 gram C] and Fluxes [GtC/-year] are large relative to Atmospheric CO_2 content and its annual increase. Large arrows for natural fluxes, small arrow fluxes arise from human activity.

1.2 CO₂ and global warming

During the past century mankind has begun to influence the geochemical system on a global scale. With respect to the C cycle there is the well documented emission of fossil fuel derived CO_2, exponentially increasing in time until about 1973, to an annual rate which is currently more or less stable at about 5-6 GtC/yr (Figure 1). Time series observations (15,16) have shown an increase in atmospheric CO_2 levels from about 270 ppm in the nineteenth century through about 315 ppm in 1958 to about 350 ppm in 1988 (Figure 3). This level is already much higher than ever before in the past 160,000 years (Figure 2). It is still increasing at an average rate of about 1.4 ppm (17) with a recent upsurge to about 2.4 ppm over the past 18 months (18). Latter surge may relate to the dramatic increase in clearing of Amazon rainforest over 1987 and beyond, but is otherwise expected to prove transitory, as also found for earlier fluctuations driven by the 1982/1983 and earlier El Niño-Southern Oscillation events.

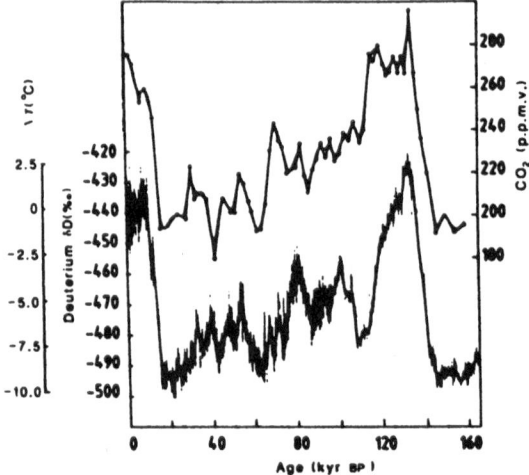

Figure 2. CO_2 concentrations ('best estimates') and smoothed values (spline function) in parts per million (volumetrically) plotted against age in the Vostok record (upper curve) and atmospheric temperature change derived from the isotopic profile (lower curve) of Deuterium. Taken from ref. 120.

Mankind has truly ventured into its 'greatest' experiment ever (19). The growing envelope of atmospheric CO_2 may, through enhanced heat absorption, lead to a global warming trend, the so-called 'greenhouse effect', possibly accompanied by melting of ice caps and a general sea level rise (20-25). Here the oceans again play a key role with respect to mostly physical processes which otherwise are beyond the scope of this paper: heat exchange with the atmosphere, shifts in ocean circulation and thermal expansion of the water column. The likelyhood of CO_2 induced global warming is very high and approaches certainty (also judging from aforementioned ice core records) but accurate prediction of the exact timing and magnitude of this event is not yet possible. Keen observers have reported an increase in global surface air temperature of about 0.5 °C over the past century (26). Yet the underlying database as well as the statistical (long vs. short term, signal vs, noise) significance is debatable (27). More importantly the suggested trend may also be ascribed to the fact that the deep ocean (with a response time of centuries) and atmosphere are probably still adjusting (19) from the 'Little Ice Age' which was most severe in the early 17th century (28).

1.3   Ocean uptake of excess CO_2

Comparison of emission rates derived from fossil fuel (coal, oil, gas) statistics (22) with the observed CO_2 increase has shown that only about half (estimates vary between 25-60 per cent) of the fossil fuel derived CO_2 is retained in the atmosphere. The other half (range 40-75 per cent) is presumably taken up by the oceans. The uncertainty derives from the fact that net CO_2 emissions due to changes in terrestrial biomass (deforestation, agriculture) are very difficult to assess. Estimates vary between extremes of -1 and +4 GtC/yr, corresponding to a minor increase or larger decrease of terrestrial biomass respectively. Nevertheless it is obvious that the oceans serve as major sink for atmospheric CO_2. Through dissolution of CO_2 in the upper mixed layer (about 75 m depth) of the ocean as well as biological C-fixation (photosynthesis) the surface ocean can take up about 2-3 GtC/yr, with an upper extreme estimate of 7 GtC/yr (Figure 1). Biological activity may well be the most crucial factor controlling the actual rate of CO_2 uptake. The net amount is only a fraction of the gross 100 GtC/yr annual atmosphere/upper ocean exchange. The strictly physico-chemical uptake of CO_2 in surface waters is further predicted to actually decrease with increasing overall CO_2 levels (Revelle factor, 29). The actual effectiveness of the net ocean sink is otherwise not well known and also varies from year to year. For example, significant changes in atmospheric CO_2 were noted during El Niño-Southern Oscillation events in the South Pacific Ocean (30-32). The exchange of CO_2 between atmosphere and ocean is clearly not in steady state but subject to dynamic variability throughout various time scales, further intensified by the recent changes generated by industrial CO_2 emissions and dramatic clearing of tropical forests.

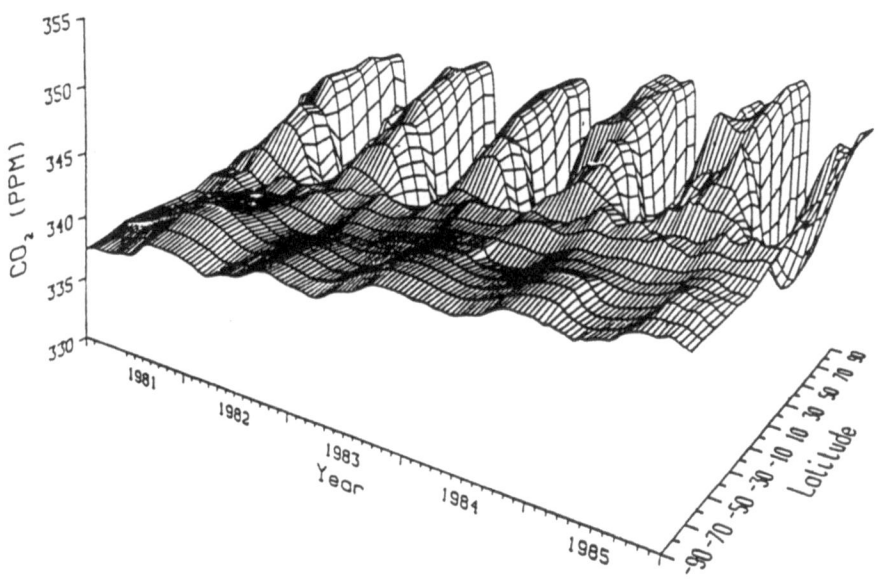

Figure 3. The increasing level of atmospheric CO_2 as a function of latitude over the 1980-1986 period. The pronounced seasonal oscillation in the Northern Hemisphere reflects the growth season of terrestrial plants, the latter being more abundant due to the predominance of landmasses in the Northern Hemisphere.

The excess CO_2, ca. 2-3 GtC/yr, taken up in the surface ocean is eventually carried downward both by water circulation (sinking of CO_2 saturated water at the poles) and biogeochemical processes. We are now, however, unable to quantify whether the physical and chemical processes alone (ocean heating, cooling, mixing, CO_2 dissolution) or the biological processes (photosynthesis and respiration) predominate the contemporary oceanic sink, both with respect to capacity as well as actual uptake rate of CO_2 (33). Clearly a better understanding and quantification of latter processes is crucial for modelling the atmospheric CO_2 content and ensuing 'greenhouse' effect. Scenarios for the latter (34-41) are in great need of proper oceanographic constraints (42) before they can serve as reliable tools for policy decisions on matters as deforestation, curtailing CO_2 emissions and anticipation of a possible sea level rise (25).

1.4   Carbon pools and pathways in the oceans

Throughout most of the world ocean there exists both a seasonal and a permanent thermocline at depth intervals typically near 50-150 m and 200-800 m respectively, which are characterized by strong temperature gradients. The corresponding density gradients largely prevent vertical mixing of water and its dissolved constituents between the surface ocean and the deep ocean interior.

Only in the surface mixed layer, the upper 25-250 m driven by wind mixing, gases like CO_2 are exchanged with the atmosphere. The direction of CO_2 gas exchange is dictated by straightforward equilibrium considerations (under or oversaturation). The actual rate of exchange is in the order of one year (43) determined by interaction of complex processes and factors such as wind induced mixing, bubble injection, surface film thickness (44).

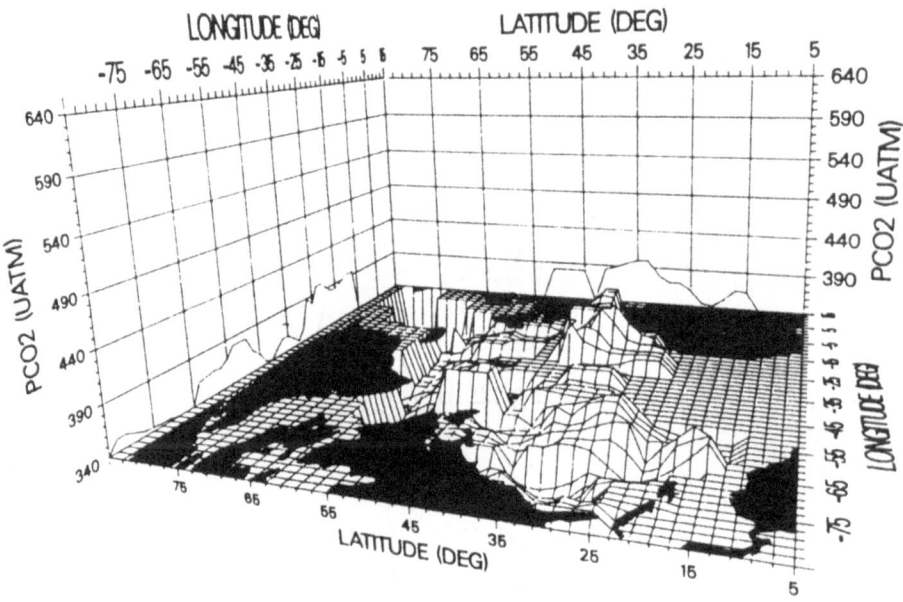

Figure 4. The partial pressure of CO_2 in surface seawater (1-15m) of the North Atlantic Ocean as measured during the TTO expeditions. This map represents a clear trend of over cq. undersaturation in equatorial cq. subpolar waters. Taken from ref. 33.

1.5 Biological production

The very same surface layer is also known as the euphotic zone, the depth interval (down to about 1% light penetration) where photosynthetic C fixation (primary production) by algae is responsible for the massive conversion of dissolved inorganic bicarbonate into plant organic matter. The process is regulated by physical forcing: light regime, upward mixing of necessary nutrient elements (N, P), temperature. Overall the C fixation leads to lower values of DIC (Dissolved Inorganic Carbon) in the surface ocean (Figure 4), so that further physicochemical dissolution of atmospheric CO_2 is possible. Much (typically 80-95%) of this Gross Primary Production is converted through grazing by zooplankton, microbial degradation and other processes within the complex food web of the upper ocean. The net results are the build-up of a pool of Dissolved Organic Carbon and replenishment of the Dissolved Inorganic Carbon pool through complete recycling (mineralization, respiration) (Figure 5).

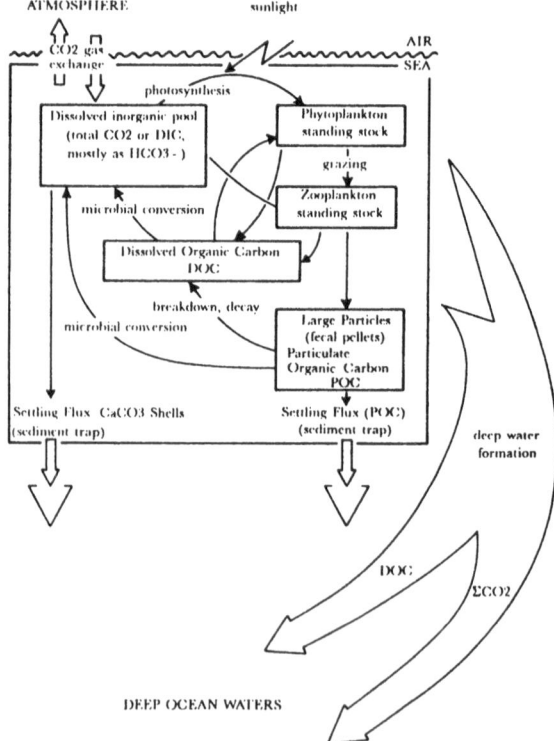

Figure 5. Simplified scheme of compartments and fluxes for C in the upper ocean. The large arrows depict the four pathways for 'pumping' Carbon into the ocean interior.

1.6 New production

However, plant/animal material or Organic Carbon (both dissolved and particulate) also escapes mineralization and is exported to deeper waters. This 'New Production' term (45-47) represents the biologically mediated net flow of C from the atmosphere into the deep ocean. New Production is typically one order of magnitude less than the Gross Primary Productivity. The larger size fractions in the particle spectrum settle down out of the euphotic zone into the deep ocean. This vertical flux of organic particles is one major pathway for transfer of atmospheric C into the deep ocean; the so called 'biological pump' for removal of atmospheric CO_2. Concurrently, biologically essential elements (N, P, Si) are also being removed to such an extent that very little remains available for plant growth. At 'near-zero' levels of nitrate, ammonia, phosphate or silicate the surface ocean productivity is typically limited by N availability. Only through replenishment of N by physical mixing from deeper waters (or possibly from atmospheric input) can the new production flux be sustained.

For the open ocean regions, which constitute 75% of the world's oceans, both the total rate of photosynthesis, and the portion of this carbon which is removed to deep waters, are uncertain to within an order of magnitude (48,49). Our limited knowledge of these crucial terms in the global carbon balance means that future atmospheric CO_2 levels, and therefore the future climate of the earth, can be predicted with little certainty. Improved accuracy of global estimates for both Primary and New Production is the objective of the ongoing Joint Global Ocean Flux Study.

1.7 Deep water formation

Another pathway for downward carbon transport is the sinking of surface water by means of intense winter cooling. The latter process occurs seasonally in the northern North Atlantic (Norwegian and Greenland Seas, Labrador Sea) and the Weddell Sea off Antarctica. Through its tendency to equilibration with the atmosphere the very cold water acts as a sink for inorganic CO_2, moreover it also carries down DOC.

Sluggish circulation within the dark abyss slowly fills the deep basins of the oceans; the mean 'age' of the deep water is about 500-1,500 years (43,50). During this time period most of the Dissolved Organic Carbon is converted to Dissolved Inorganic Carbon. The settling large particles are also remineralized. From the vertical and interoceanic distribution of various dissolved tracers (nitrate, phosphate, ΣCO_2, ^{13}C, etc.) in the world ocean we know that the net flux of organic matter originating from the surface ocean is remineralized within the deep sea and at the sea-floor. We are not certain whether this organic matter is brought into the deep sea largely as particles settling from the euphotic zone, or mostly as Dissolved Organic Carbon carried along with the general water circulation.

The water balance of the surface ocean is restored by a mean upward advective flow or upwelling in the order of 4 m per year. In certain regions, e.g., in the equatorial Atlantic, more intense upwelling is noted from excess CO_2 contents of the water as compared with the atmosphere (43, see also Figure 5). Some of this excess is fixed by photo-synthesis, some escapes into the atmosphere, as was conclu-ded from $^{13}C/^{12}C$ analyses of atmospheric CO_2 (51). The short mixing time of the atmosphere makes each hemisphere fairly homogeneous with respect to CO_2; the ocean 'breathes' CO_2 by inhaling at the poles and exhaling at the equator. This ocean/atmosphere CO_2 exchange is massive (about 100 GtC/a) compared with the modern fossil fuel CO_2 additions (about 4 GtC/a) (Figure 1).

1.8 Vertical transport of particulate organic matter

The settling of large particles (52) has been inves-tigated extensively in the past decade with sediment traps (53-57). Compilations of Pacific data suggest that more than 75% of the particle flux is recycled within the upper 500m (56,58). In the deep ocean vertical gradients of the fluxes of organic matter (C/N/P, organic compound classes, etc.) point at significant microbial mineralization and transformation during settling (59-62). From the observed seasonality of long term records we now also know that the settling velocity can be very rapid indeed, effectively bringing the C down to the seafloor within about one month after a plankton bloom (63).

1.9 Dissolved organic matter

The alternative route for bringing biologically fixed C to the deep ocean is through Dissolved Organic Carbon being carried down with the sinking water masses in subpo-lar regions. Recently published results strongly suggest this route (64). The undersaturation of oxygen or Apparent Oxygen Utilization (AOU) in the deep ocean would be largely accounted for by mineralization of Dissolved Organic Car-bon. Another implication would be the significance of free ranging bacteria, rather than those attached to settling particles (65,66). These relations and the underlying data for Dissolved Organic Carbon are currently the subject of considerable debate (67,68) and investigations.

1.10 The microbial loop

The omnipresence and abundance of bacteria has received considerable attention in the past decade (69-72).
Relationship between bacterial biomass and chlorophyll (73)
or primary productivity has been established within a range
of marine systems. Most of the bacteria appear to be free
living (rather than attached), utilizing Dissolved Organic
Carbon rather than particles as a food source (69,74).
Through such studies the concept of a substantial microbial
foodweb has been advanced, where the bacteria are sustained
by a flow of Dissolved Organic Carbon derived from phytoplankton and herbivores. The dissipation of C through this
passway has now also been demonstrated in a large scale
field study (75). Within the euphotic zone the overall effect of the bacteria may be net remineralization (76), enhancing the efficiency by retaining essential nutrients
within the surface ocean (77-78). The ensuing wax and wane
of phytoplankton blooms (79) would be closely followed by
changes in the flux-rate of CO_2. The conventional description of flows of energy and matter through trophic levels
(phytoplankton, herbivores, carnivores, settling flux of
fecal pellets) is now augmented by microbial pathways (Figure 5). For the deep ocean we now have evidence that free
ranging bacteria play a major role in mineralization in the
ocean interior as well (65,66).

These insights somewhat parallel the shift in thinking
towards the cycling of Dissolved Organic Carbon as a means
of describing C fluxes in the sea, at the expense of the
particle flux concept. Above we mentioned substantial evidence for both routes for transfer of C into the deep sea.

1.11 Natural pathways for pumping carbon into the oceans

So far we dealt with the dissolution of CO_2 in seawater and its uptake as POC and DOC into particulate and dissolved organic matter. In addition Coccolithophorideae,
planktonic Foraminifera and planktonic Ostracods also extract CO_2 (about 3 GtC per year) from surface seawater in
the form of $CaCO_3$ shells. Much of this material eventually
settles to the seafloor and thus constitutes another flux
of C into the deep sea (53,80).

In summary we now know four (rather than three; 81)
different routes currently used by the geosphere-biosphere
system for 'pumping' C into the deep ocean:
1. Dissolved Inorganic Carbon carried down during deep water formation (DIC)
2. Dissolved Organic Carbon carried down during deep water
   formation (DOC)
3. Vertical flux of organic matter in settling particles
   (POC)

4. Vertical flux (sedimentation) of calcite shells ($CaCO_3$)
At present we cannot say which routes are most important.
Ongoing worldwide research programmes like the Joint Global
Ocean Flux Study are focusing on exactly this problem.

Within the deep ocean the organic matter (2. and 3.)
is almost all converted again by microbes and animal respiration to dissolved CO_2. Also in the deepest parts of the
ocean the settling calcite shells (4.) tend to dissolve
again. For deeper and older water the total CO_2 content
(ΣCO_2) is therefore larger, for example about 2,200 and
2,400 micromoles per liter (μM) in deep (4 km) Atlantic and
Pacific water respectively, as opposed to values in the
order of 2,000 μM in surface water. Latter surface water
values vary considerably as a result of the effect of temperature and salinity on solubility, as well as due to biological uptake.

1.12 CO_2 in seawater and dissolution of calcite

CO_2 in seawater is involved in a complex system of
equilibria (82), given below in a simplified manner. At the
sea surface CO_2 tends to dissolve in accordance with Hen-
ry's Law:

$$[CO_2] \text{dissolved} = \alpha \cdot pCO_2 \text{gas} \qquad (1)$$

where α is a function of salinity and temperature. Upon
reaction with water

$$CO_2 + H_2O <=> H^+ + HCO_3^- \qquad (2)$$

some of the thus formed bicarbonate further dissociates

$$HCO_3^- <=> H^+ + CO_3^{2-} \qquad (3)$$

into another proton (H^+) and the carbonate ion. Reactions
(2) and (3) are governed by the equilibrium dissociation
constants

$$K1 = [HCO_3^-][H^+]/[CO_2] \qquad (4)$$
and
$$K2 = [CO_3^{2-}][H^+]/[HCO_3^-] \qquad (5)$$

Both K1 and K2 are functions of temperature, salt content,
pressure, etc., in general to such an extent that at lower
temperature the seawater contains more CO_2. The total
amount of CO_2 or ΣCO_2 consists of

$$\Sigma CO_2 = [CO_2] + [H_2CO_3] + [HCO_3^-] + [CO_3^{2-}] \qquad (6)$$

where in seawater about 90 per cent is bicarbonate [HCO_3^-], about ten per cent carbonate $[CO_3^{2-}]$, and very little $[CO_2]$.

Finally we have to take into consideration that the two anions $[HCO_3^-]$ and $[CO_3^{2-}]$ have an electrical charge, which has to be balanced by the charge difference between all cations (Na^+, K^+, Mg^{2+}, Ca^{2+}, etc) and other anions (mostly Cl^- and SO_4^{2-}). Latter charge difference or Alkalinity is more or less constant in seawater and thus matches the sum of charges of the CO_2 system:

$$Alkalinity = [HCO_3^-] + 2[CO_3^{2-}] \tag{7}$$

where Alk is expressed as charge equivalents per liter. Within seawater the six variables ($[CO_2]$, $[HCO_3^-]$, $[CO_3^{2-}]$, $[H^+]$, Alk, ΣCO_2) of the carbonate system are governed by latter four relations (4-7), which also determine the $[H^+]$, i.e., the acidity or pH. By measuring two variables, e.g., ΣCO_2 and Alk, one may calculate the other four. In this way we may for example assess $[CO_2]$ in surface water and through above (1) calculate the predicted pCO_2 in surface water. Latter value more often than not is above or below the pCO_2 in the atmosphere, i.e., there is no equilibrium distribution between atmosphere and upper ocean. For example most cold Northern waters are undersaturated, whereas warm equatorial waters are generally oversaturated (Figure 4). Fixation of CO_2 by plankton growth also leads to undersaturation.

Addition of CO_2 to seawater does take place in the deep sea by respiration of dissolved and particulate organic matter:

$$Organic\ Matter\ (HCOH) + O_2 = CO_2 + H_2O \tag{8}$$

and yields a higher value for ΣCO_2 which then causes shifts in all the above equilibria. Curiously enough the higher CO_2 content eventually leads to a lower $[CO_3^{2-}]$ concentration. This is understood by looking at (7) where at higher ΣCO_2 (6) the constant Alk can only be accommodated by shifting into direction of lesser charged $[HCO_3^-]$ ion. This is exactly what happens in the deep ocean and deeper or older water has a higher ΣCO_2 but lower $[CO_3^{2-}]$ concentration. At greater depth this concentration is below the solubility product of Calciumcarbonate

$$S.P.\ (CaCO_3) = [Ca^{2+}][CO_3^{2-}] \tag{9}$$

and the calcite shells tend to dissolve. The well documented $CaCO_3$ distribution in the surface layer of marine sediments (43,83) indeed shows a distinct horizon above which $CaCO_3$ is well preserved whereas virtually no $CaCO_3$ is found in sediments below that horizon. With increasing age, i.e.,

increasing CO_2 content and lower $[CO_3^{2-}]$, of the deep water this horizon shoals from depths around 3.5 km in the
North Atlantic to about 1.2 km in the North Pacific Ocean.

2. STORAGE OF CO_2 IN THE OCEANS

2.1 Uptake in the surface ocean

For calculating various greenhouse effect scenarios
the most popular assumption is a doubling of atmospheric
CO_2 as the forcing function. The nowadays 350 ppm is compared with 700 ppm at doomesday. Using above equations (1-
7) we simply calculate the effect for surface water at 10 C
and salt content of 35 per mille and typical Alkalinity of
2,000 micro equivalent per liter:

Temp	S	Alk	pCO$_2$	ΣCO$_2$	[HCO$_3^-$]	[CO$_3^{2-}$]	pH	R.F.
10	35	2,000	350	1,882	1,732	134	8.15	10
10	35	2,000	700	1,964	1,866	67	7.88	17

This assumes that the surface ocean on average is in equilibrium with the atmosphere. In each case we then calculated (not shown) the effect of a small change in pCO_2 on the
change in ΣCO_2 :

$$\text{Revelle Factor} = \frac{\text{Per cent change of } pCO_2}{\text{Per cent change of } \Sigma CO_2} \qquad (10)$$

and found the R.F. to increase from about 10 to about 17.
In other words the inorganic buffering capacity of the
oceans is predicted to decrease. For colder or warmer surface waters (the modern surface temperatures range from
about -1 to 30 °C) values for ΣCO_2 as well as R.F. would be
higher or lower, yet in all cases one would find the buffering capacity to be lower at higher atmospheric pCO_2.
Assuming a constant fossil fuel CO_2 production of about 5
GtC/year one would expect that less and less can be accommodated in the oceans. Where currently still about half or
2-3 GtC/yr is taken up, this may drop to 25% or less at
higher atmospheric pCO_2. This is a positive feedback with
detrimental effect: the atmospheric inventory and corresponding pCO_2 would rise more and more rapidly than the current rate of 1.4 x 10-6 atm/year. One important caveat here
is that the normal biological C fixation of the oceans is
left out. The more important latter process is in the modern ocean, the lesser we have to worry about decreasing
inorganic buffering capacity. Above we mentioned several
lines of evidence suggesting that the role of the oceanic
biota, i.e., the importance of the 'biological pump', may
well have been underestimated. Models emphasizing the role

of biological processes are being developed (84,85) and trial runs hint at the key role of biota (84). More accurate quantification of biological versus inorganic C uptake by the oceans is also a main objective of the Joint Global Ocean Flux Study which integrates field observation programmes with modelling efforts.

Another important observation is that the $[CO_3^{2-}]$ concentration in ocean waters will undoubtedly decrease at higher atmospheric pCO_2. Also the drop in pH represents a more acidic environment, with numerous possible implications for the biota, for example shifts will take place in the chemical state (hence toxicity) of trace metals in surface waters.

2.2 On the potential for buffering by calcite

Upon sinking into the deep ocean the newly formed deep water will in the future contain more and more CO_2, hence have a lower $[CO_3^{2-}]$ concentration. The subsequent conversion of organic matter (see above) within the deep waters will be rather similar in both the current and the doomesday ocean, i.e., in both cases an extra amount of CO_2 will be added to ΣCO_2 and the $[CO_3^{2-}]$ will drop accordingly. Overall the deep water of the future ocean will have a lower and lower $[CO_3^{2-}]$ concentration. This now would lead to more intense calcite undersaturation, i.e., more extensive dissolution of $CaCO_3$ from surface sediments:

$$CaCO_3 = Ca^{2+} + CO_3^{2-} \qquad (11)$$

where concurrent shifts in the equilibria (4-7) would yield readjustment, notably of pH.

It might be argued that overall the addition of CO_2 from fossil fuels will thus simply be buffered by dissolution

$$CaCO_3 + H_2O + CO_2 = Ca^{2+} + 2\ HCO_3^- \qquad (12)$$

of calcite sediments. With a known reservoir of fossil fuels of 7,000 to 10,000 GtC (of which about 1,000 GtC as petroleum) it appears that the equivalent of CO_2 can easily be compensated for by dissolution of the same 7,000-10,000 GtC of calcite, only a tiny fraction of the total 60,000,000 GtC in calcite sediments (Figure 1). Eventually a new quasi equilibrium state would be reached with the CO_2 content of the atmosphere still about twice that of today (600 ppm; 86,87) and higher total CO_2 content (e.g., some 20,000 GtC above the current 40,000 GtC) and Ca^{2+} content of ocean waters. This doubling of the atmospheric CO_2 is still troublesome but otherwise the most optimistic scenario. This mass balance or budget approach is in itself not

158

incorrect. However, in order to be effective it requires a
very well mixed atmosphere/ocean/sediment system in order
to accommodate reaction rates as fast as the rate at which
we are adding fossil fuel CO_2 to the atmosphere.

2.3 Dynamics of the real ocean system

The key to the CO_2 greenhouse problem lies in the
turnover rates in the geosphere - biosphere system (Figure
6). Upon introduction into the atmosphere from a car ex-

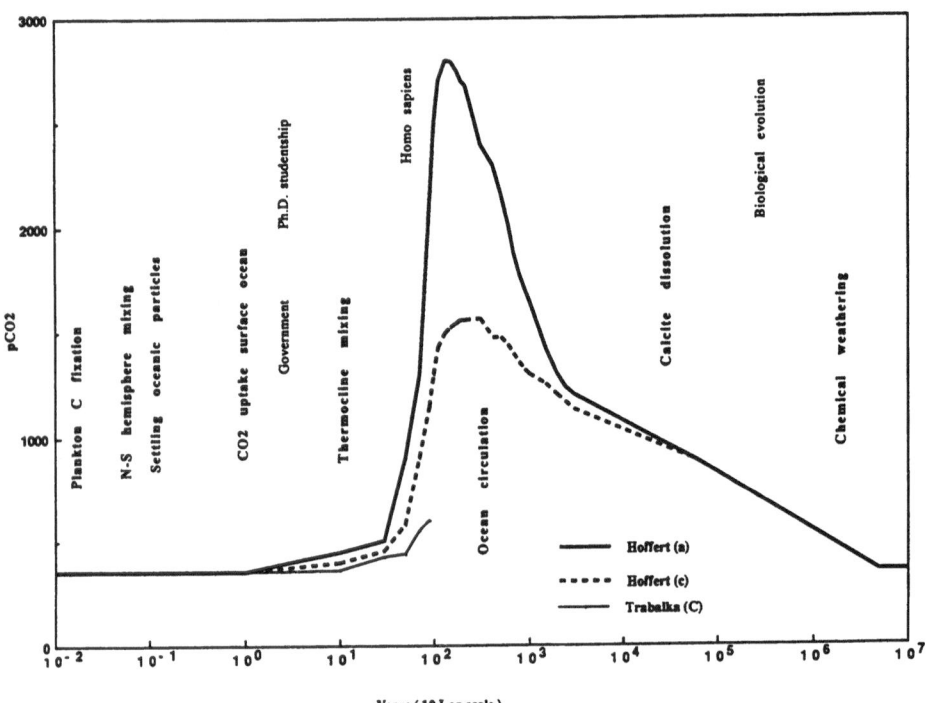

Years (10 Log scale)

Figure 6. Sketch of the typical time response of the geosphere-biosphere to the rapid emission of fossil fuel CO_2.
Logarithmic horizontal axis covers responses over the range
from one month (0.1 year) to 10 million (10^7) years. Vertical axis for the predicted atmospheric CO_2 concentration in
parts per million [ppm] or 10-6 atm units. The curves (a)
and (c) of Hoffert c.s., as also shown in Figure 8. The low
emission scenario (C) of TRABALKA c.s., (ref., 91) as in-
spired by Lovins c.s., All scenarios predict a peak of CO_2
within the next 100-200 years. Biology cannot adapt itself
(evolution) so very rapidly. The short life time of our
governments is on the other hand unfavourable for effective
decision-making on long term problems.

haust or smoke stack the CO_2 within the Northern hemisphere is homogenized within timescales of days or weeks. Mixing of air between the Northern and Southern hemisphere takes several months to a year, hence the pCO_2 in the Southern hemisphere lags slightly behind that in the more industrialized Northern hemisphere. Surface waters of the ocean need about one year to equilibrate their CO_2 content with that of the atmosphere. It takes several decades for this surface water signal to penetrate into the underlying several hundred meters of the thermocline region. The real formation of deep water (by sinking in polar regions) and downward transport of associated CO_2 takes several centuries to about 1,000 years. The majority of ocean water has not been at the surface for more than 500 years, i.e., it was already submerged before CO_2 was discovered, Galileo entered into a trivial debate or Columbus sailed to the Americas.

Only a very small portion of the huge $CaCO_3$ reservoir is exposed at the surface of deepsea sediments available for, in itself also slow, dissolution. The response time of this final dissolution step is estimated to be at least 10,000 years, if not much longer (88). Over even longer, geological, time scales of millions of years the ocean/ atmosphere C budget may be further readjusted by shifts in the chemical weathering of continental rocks (Figure 6), as also happened in the past (89). Higher atmospheric CO_2 levels lead to increased chemical weathering which removes CO_2 from the atmosphere, a negative feedback loop which, however, acts extremely slowly. Evolutionary adaptation of the biosphere would also operate in the 100,000 to million year time scale, i.e., the current rate of change will lead to catastrophic extinctions and an overall much lower species diversity. Otherwise such long time scales are irrelevant for supporting policy decisions with time scales ranging from 4 years (life-time of a government or a graduate studentship) to a century (life span of Homo sapiens). Similarly for predictive models of the greenhouse effect in the coming 5,000 - 10,000 years the calcite buffering can also be neglected altogether.

The marine biological pathways have a more promising response. The turnover rate of plankton in the surface ocean is in the order of days (as opposed to about ten years for land plants). The overall marine biomass of about 4 GtC may be modest compared to land plants (600 GtC), the annual marine turnover in the order of 50-100 GtC/year (gross primary production) is significant. From all C fixed by marine photsynthesis about ten per cent settles out of the euphotic zone into the deep sea. Within one month after a plankton bloom event such material arrives at the seafloor. Better assessment of marine primary carbon fixation and the ensuing settling flux of C into the deep sea is another focal point of the JGOFS study.

Various models based on these typical time constants generally will, for a given fossil fuel CO_2 input, predict the oceans to continue slowing down the atmospheric CO_2 buildup in the first several hundred or thousand of years (43,90,91).

Assume a constant input of 5 GtC/year of which currently 3 GtC/yr is taken up by the oceans. If latter uptake is dominated by biological processes than it may well remain at about 3 GtC/yr, if dominated by chemical dissolution then the ocean uptake will be sharply reduced in the future. For policy decisions on curtailing CO_2 emissions, accurate predictive CO_2 models over the decade to century time scale are needed. Improved accuracy of such predictive models of atmospheric CO_2 can only be achieved by more accurate quantification of biological versus chemical CO_2 uptake by the oceans.

Adjustment of above time constants will yield a different dynamic response within the critical 10 - 1000 year time domain. Yet eventually after, perhaps 5,000 years all such models will converge to very similar final steady state values for atmospheric pCO_2 and CO_2 stored in ocean waters.

2.4 Technical fixes for temporary storage of CO_2 in the deep sea

The immense prospect of a CO_2 greenhouse effect has given rise to very many imaginative schemes for resolving the problem (92). The most appealing and feasible solutions rely on CO_2 emission reduction by both lowering energy consumption as well as shifting to alternative energy sources. Other schemes investigate storage in biomass on land (93,94) or increasing marine biomass uptake (95), both requiring huge fertilization (N, P) programmes and less attractive also for other reasons. Here we shall review the various options for storage of CO_2 in the deep ocean:

Marchetti proposed collecting of CO_2 effluent at large point sources (power plants), transporting by pipeline and disposal by injection of CO_2 into natural deep sea currents, notably the Mediteranean Outflow Water (96, 97). In fact, large point sources like power plants account for less than one-third of total fossil fuel related CO_2 flux (the remainder largely diffuse sources, notably traffic). Shifting to other energy carriers (H_2, NH_3) for traffic and transport has been suggested as to further centralize the CO_2 emission points beyond the current 30 per cent (96). Otherwise the assessment by Marchetti of cost and technical problems is grossly optimistic and hypothetical compared to below assessments of Baes c.s., and Steinberg c.s., (99-110). Also 30% of all West European CO_2 production amounts to only a few per cent of the worldwide CO_2 production. Finally the Mediterranean Outflow into the North Atlantic does indeed beyond Gibraltar sink to depths around 1,200 m. However, its distribution and mixing pattern within the North Atlantic and beyond warrants large scale atmosphere/ocean ventilation of MOW within decades after leaving Gibraltar, notably in the Equatorial Atlantic region and along the coast of West Africa. At best about half of the thus injected CO_2 would be released into the atmosphere some 50-200 years later.

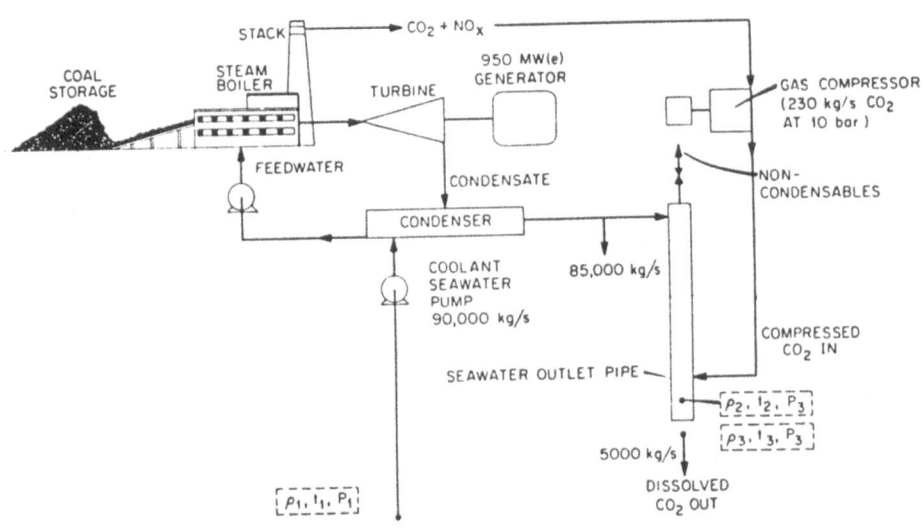

Figure 7. Flowsheet for a floating coal-fired power plant with CO_2 disposal to the deep (1,000-3,000m ocean). Coal would be transported towards the plant with barges, electricity transferred to land through cables. Taken from Baes c.s., (101).

Similar injection into the Iceland-Greenland (Denmark Strait) and Iceland - Scotland Overflow deep currents (98) would contribute to the southward flowing North Atlantic Deep Water (NADW) complex. Part of this complex is the Western Boundary Under Current (WBUC) flowing southward at great depth along the eastern seaboard of the USA. The WBUC is another deep sea current suitable for CO_2 injection. Part of this NADW or WBUC would, due to equatorial Atlantic upwelling, contact the atmosphere again, yet a large portion will remain at a depth flowing southward towards the

Antarctic Ocean. A substantial portion will indeed continue to flow around the Antarctic into the deep Pacific and Indian Oceans and will remain submerged for a period up to 2,000 years. Yet gradually more and more of this water will also emerge at the surface and release its excess CO_2. The World Ocean Circulation Experiment now starting will provide us with more reliable estimates of the time scales involved, yet on average the NADW will need several decades before reaching the surface again. Any CO_2 thus injected will be released into the atmosphere with a delay in the order of 200-500 years.

The Oak Ridge National Laboratory group around <u>Charley F. Baes</u> produced a series of papers (99-102) devicing several schemes all based on initial collection by scrubbing power plant stacks with monoethanolamine (MEA) and transport of compressed (150 atm) CO_2 in pipelines. Large US power plants account for only 30% of US CO_2 output, or 10% of global CO_2 emissions. Baes c.s., estimate that the extraction, compression to 150 atm and transport over 40 km of CO_2 requires half of the energy produced, or only 1/3 if the coal was burned in pure oxygen from an air separation plant (including the expenditure for running latter plant). Alternatively a plan was deviced for building the whole power plant on a barge floating over the deep ocean, transporting coal/oil and electricity to and from the barge. Some 36% of the electrical output would be needed overall to run the barge plant itself. Several options for final disposal in the oceans are suggested:

a) High density solution in seawater, the reject stream would sink to the ocean floor
b) Pure liquified CO_2 needs to be emitted at depths greater than 3,000m in order to be negatively buoyant (i.e., sinking) relative to ambient seawater. A large pool of liquified CO_2 would remain dormant on the seafloor
c) Dumping of solidified blocks of CO_2-hydrate. The conversion to CO_2-hydrate requires considerable additional energy. The blocks would sink to the seafloor and remain there
d) Dumping of blocks of solidified CO_2 (dry ice). The conversion to dry ice requires considerable additional energy. The blocks would sink to the seafloor and remain there

Preferably all four physical states (phases) in discharge options a) through d) would provide long term storage on the deep ocean floor. However, the passing ocean currents, which are undersaturated in CO_2 with respect to all these four phases, would erode the CO_2 piles, carrying along the erosion product in the form of total dissolved CO_2 in seawater. Eventually the deep water would reach the surface again and the CO_2 would escape into the atmosphere.

In conclusion all schemes of Baes et al., would lead to a delay of several hundreds of years, as further worked out belowin the discussion of Hoffert's c.s., and Lysen's studies (103,104).

Steinberg and his associates produced an exhaustive series of papers on solutions to the CO_2 problem, some of them beyond imagination (105-110). For example conversion of CO_2 to marketable products yields a consumption of at best only 1-2 % of CO_2 production. Also these marketable products would within very short time span again be released as CO_2. Their specific scheme on deep ocean injection (106) appears least unattractive and otherwise closely resembles design, specifications and efficiency of the Baes et al., systems.

Marland (92) summarizes that the first two steps of all the above-mentioned schemes, collection and transport, have been well investigated and are in fact in full operation on modest scales as commercial enterprises for the use of CO_2 in increasing recovery of petroleum from geological reservoirs (111,112). The most economical and practical methods, amine scrubbing and long distance transport as liquified CO_2, are operational today and well understood also in terms of efficiency. Some 32-45% of the combustion energy of coal would be required. Recent developments suggest more efficient operation of the MEA extraction step (113) requiring only about 15-25% of the overall energy input. This combined with the expenditure for compression, transport and the like would make the overall energy loss somewhat lower, around 20-35 per cent. The disposal of CO_2 into the deep sea is less well documented and might also require additional energy. For a more accurate assessment of technical feasibility and calamity risks one might also look at the (relatively shallow water, <1,000m) installations of the offshore petroleum industry.

2.5 Fate of CO_2 injected into the deep ocean

Hoffert and associates tested the various deep injection scenarios with a simple box diffusion model (103). The input of 7,000 GtC over an initial 150 years time period is injected into the ocean in five different modes:
a) 100% into the atmosphere, none directly into the ocean
b) 50% into the atmosphere and 50% into the ocean at 1,500 m depth
c) 50% into the atmosphere and 50% into the ocean at 4,000 m depth
d) 100% at 1,500 m depth
e) 100% at 4,000 m depth
The outcome of the modelling as shown in Figure 8 further

164

emphasizes the buffer function of the oceans. The first
option (no injection) is virtually identical to aforementioned greenhouse scenario of Siegenthaler & Oeschger (90).
By deep injection the thermocline barrier between surface
ocean and deep ocean is effectively bypassed and in extreme
option e) the normal renewal time of deep water of about
1,000 years is thus avoided when bringing CO_2 into the deep
ocean. However, the same deep water renewal then also
brings the injected CO_2 back to the surface with the same
1,000 year mixing time and eventually after about 5,000
years all five options converge to the same steady state
end value of about 1,150 ppm CO_2 in the atmosphere,

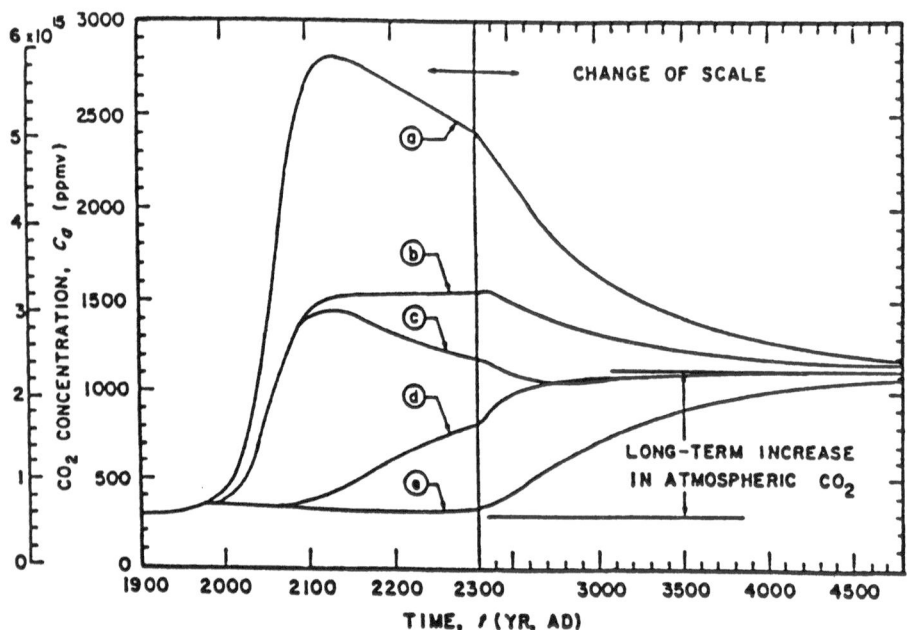

Figure 8. Model projections of atmospheric carbon dioxide
variations assuming entire fossil fuel CO_2 reserve is depleted as a logistic function of time: a) 100% injected
into the atmosphere; b) 50% injected at 1,500 m depth into
the ocean; c) 50% injected at 4,000 m depth into the ocean;
d) 100% injected at 1,500 m depth into the ocean; 100% injected at 4,000 m depth into the ocean. NOTE fivefold compression of time-scale beyond 2300 A.D. Taken from Hoffert
c.s., (103).

roughly four times higher than the pre-industrial 265±10
ppm level before about 1860 (114).
    We mentioned earlier that large stationary power
plants account for at most 30% of the current CO_2 produc-

tion, i.e., the maximum effect of deep injection would lie somewhere in between option a and options b. and c. Some 20-45% of the power output of such plants would be lost in the process, i.e., society would still have to reduce its electricity use accordingly. The peak value of the transient response in the year 2100 A.D. (Figure 8) would exceed the 1,500 ppm for scenario b or c.

Finally it should be noted that the Siegenthaler & Oeschger scenario (90) for production, emission and fate of CO_2 as used by Hoffert c.s., is among those with the highest CO_2 input function. Most recent input predictions are considerably lower (91) and the resulting atmospheric CO_2 levels would also be much lower. For example Trabalka et al., (91) would in 2075 A.D. arrive at about 1,600, 600 and 450 ppm for a suite of high, standard and low production scenarios, all relying on 100% atmospheric emission.

Lysen was also invited to review the deep sea disposal option (104). He observed that the input function of 7,000 GtC in 150 years of Hoffert c.s., (104) might in fact be lower, partly as a result of policy measures towards reducing the burning of fossil fuels. He only assessed the final (5,000 years) steady state value as function of an integrated CO_2 input ranging from 200 to 7,000 GtC., using a simplified time independent version of the Hoffert c.s., model. For example with a past input of about 400 GtC (170 GtC from fossil fuel, 115; upper limit of 150 GtC from deforestation) and a future input of 1,000 GtC the integrated 1,400 GtC adition to the ocean/atmosphere system yields a steady state atmospheric CO_2 level of about 400 ppm 5,000 years from now.

The transient response has not been calculated, but it appears reasonable to assume that the peak value in the year 2100 A.D. will be reduced in roughly a proportional manner. When glancing at (Figure 8) but again realizing that at best 30% (stationary plants) of CO_2 production can be injected into the deep sea the peak atmospheric CO_2 value in 2100 A.D. would probably still exceed 500 ppm, roughly the typical doubling of atmospheric CO_2 commonly used as input function in General Circulation Models (GCM) for predicting global temperature (climate) response.

Upon one remarkable but not very well supported suggestion of Marchetti (96,97) several investigators were challenged to make an effort towards serious initial assessment of feasibility (99-110). Others were faced with the task of putting all this in perspective (92, 104, this review). From the combined theoretical studies it now appears that at best deep sea injection of at most 30% of all fossil fuel CO_2 is feasible, at the expense of some 20-45% of the generated electricity. Otherwise it is and has never been considered a serious practical option (116,117). Only when other options (energy conservation, other energy sources, to mention a few) are even less attractive some further study and pilot plant experiment would be warranted.

3. CONCLUSIONS

1. The global Carbon cycle is dominated by the oceans.

2. From the time integrated fossil fuel CO_2 emission in the past about half has been taken up by the oceans.

3. From the current annual fossil fuel CO_2 emission of 4-5 GtC/year at least 2 GtC/year is taken up by the oceans.

4. The oceanic uptake occurs through both strictly inorganic (physico-chemical) dissolution as well as through various biological processes. Present global estimates of both routes have an uncertainty within an order of magnitude. Currently we are not sure whether the abiotic or the biotic uptake route dominates.

5. The abiotic oceanic uptake will definitely decrease with increasing CO_2 content of the ocean/atmosphere system. The biotic oceanic uptake is more likely to remain constant or increase.

6. More accurate quantification of the biotic and abiotic uptake routes in the modern ocean is crucial for predicting future levels of atmospheric CO_2.

7. Collection, transport and deep ocean disposal of CO_2 is only conceivable for large stationary plants, at most representing 30% of current CO_2 emissions. Furthermore this requires at least some 30-45% of the generated electricity and the electricity consumption of society would have to be reduced accordingly. Development of more efficient CO_2 collection methods might be possible in the future.

8. Deep ocean disposal will only delay the transient buildup of atmospheric CO_2 over decade to century time scales. The final atmospheric CO_2 level after some 1,000 or more years will not be reduced. The transient peak value for atmospheric CO_2 in about 2100 A.D. as predicted in several C cycle models might be reduced by about 50 per cent.

9. For a direct atmospheric input of 7,000 GtC within 150 years the transient peak value at 2100 A.D. would be about 2,800 ppm, for the maximum deep sea injection case of 30% the peak value would still exceed 1,500 ppm. The final steady state level in 7000 A.D. would in both cases be the same at about 1,150 ppm.

10. By strongly curtailing the modern fossil fuel CO_2 emission trend it might be possible to reduce the integrated input to about 1,400 GtC. In such case the peak value in 2100 A.D. might well be reduced to about 1,300 ppm for direct atmospheric injection or about 500 ppm for the maximum deep sea injection scenario. The final steady state value after 5,000 years would in both cases lie around 400 ppm, well above the preindustrial 260 ppm.

4. RECOMMENDATIONS

1. Research towards better quantification of the biotic and abiotic uptake mechanisms of the ocean in order to allow accurate prediction of future atmospheric CO_2 level as function of future fossil fuel input by mankind. Latter prediction would provide a safe ceiling for future CO_2 production as benchmark for policy towards curtailing predicted CO_2 production.

2. Deep sea injection is at best an only partial, energy consuming and temporary solution to the CO_2 problem. It might conceivably at best account for 30% of fossil fuel CO_2 emissions, meanwhile requiring 20-45% reduction in electricity consumption. The injected CO_2 will not be stored on the seafloor but emerge again after several decades or centuries, partly depending on injection depth.

3. Large scale reduction of current and future fossil fuel CO_2 production is the preferable option. Improved efficiency of the whole spectrum of current energy technologies is feasible (118).
   Major reduction of CO_2 emissions from the Western industrialized countries is possible, with admittedly adjusted life style but without affecting overall quality of life. The real challenge lies in devising universally acceptable schemes for curtailing the predicted exponential growth of CO_2 emissions by countries with currently developing economies (notably in SE Asia; 119).

5. REFERENCES

1. NEFTEL, A., H. OESCHGER, T. STAFFELBACH and B. STAUF-
    FER. (1988). "CO_2 record in the Byrd ice core, 50,000
    - 5,000 years BP", in Nature, 331, pp. 609-611.
2. KNOX, F., and M.B. McELROY. (1984). "Changes in atmos-
    pheric CO_2: influence of the marine biota at high la-
    titude." in J. Geophys. Res., 89, pp. 4629-4637.
3. BROECKER, W.S. (1982). "Glacial to interglacial chan-
    ges in ocean chemistry." in Progress in Oceanography,
    11, pp. 151-197.
4. BROECKER, W.S. (1987). "Unpleasant surprises in the
    greenhouse?" in Nature, 238, pp. 123-126.
5. SARMIENTO, J.L., and J.R. TOGGWEILER. (1984). "A new
    model for the role of the ocean in determining atmos-
    pheric pCO_2." in Nature, 308, pp. 624-
6. MIX, A.C., and R.G. FAIRBANKS. (1985). "North Atlantic
    surface-ocean control of Pleistocene deep-ocean cir-
    culation." in Earth Planet. Sci. Lett., 73, pp.
    231-243.
7. BOYLE, E.A. (1986). "Paired carbon and cadmium isotope
    data in benthic foraminifera: Implications for changes
    in oceanic phosphorus, oceanic circulation, and atmos-
    pheric carbon dioxide." in Geochim. Cosmochim. Acta,
    50, pp. 265-276.
8. BOYLE, E.A. (1988). "Vertical oceanic nutrient frac-
    tionation and glacial / interglacial CO_2 cycles." in
    Nature, 331, pp. 55-56.
9. SUNDQUIST, E.T., & W.S. BROECKER. (Eds.,). (1985).
    "The Carbon Cycle and Atmospheric CO_2: Natural Varia-
    tions Archean to Present." in AGU Geophys. Monograph
    Series, Washington D.C., Vol.32, 627pp.
10. MOORE, B., and B. BOLIN. (1986). "The oceans, CO_2 and
    global climate change." in Oceanus, 29, pp. 9-15.
11. BOLIN, B., BoR. DOOS, J. JAGER and R.A. WARRICK.
    (1986). "The greenhouse effect, climatic change and
    ecosystems." in SCOPE-29, John Wiley and Sons, Chi-
    chester, 541pp.
12. BROECKER, W.S., & T.H. PENG. (1987). "The role of
    CaCO3 in glacial to interglacial atmospheric CO_2 chan-
    ge." in Global Biogeochemical Cycles, 1, pp. 15-29.
13. SARNTHEIN, M., K. WINN, J.C. DUPLESSY and M.R. FONTUG-
    NE. (1988). "Global variations of surface ocean pro-
    ductivity in low and mid latitudes: influence on CO_2
    reservoirs of the deep ocean and atmosphere during the
    last 21,000 years." in Paleooceanography, 3(3), pp.
    361-399.
14. BERGER, W.H., V.S. SMETACEK and G. WEFER. (Eds.,).
    (1988). "Productivity of the Ocean: Past and Present."
    Dahlem Konferenzen, April 1988.John Wiley and Sons
    Ltd., (Chichester); in press.

15. KEELING, C.D., R.B. BACASTOW, A.E. BAINBRIDGE, C.A. EKDAHL Jr., P.R. GUNTHER, L.S. WATERMAN AND J.F.S. CHIN. (1976a). "Atmospheric carbon dioxide variations at Mauna Lao Observatory," in Tellus, V28, pp. 538-551.

16. KEELING, C.D., J.A. ADAMS Jr., C.A. EKDAHL and P.R. GUNTHER. (1976b). "Atmospheric carbon dioxide variations at the South Pole." in Tellus, V28, pp. 552-564.

17. GAMMON, R.H., E.T. SUNDQUIST and P.J. FRASER. (1985). "History of carbon dioxide in the atmosphere." In: J.R. Trabalka (Ed.,) Atmospheric Carbon Dioxide and the Global Carbon Cycle, Oak Ridge National Laboratory, Oak Ridge, U.S. DOE/ER-0239, pp. 25-62.

18. HOUGHTON, R. A. and G.M. WOODWELL. (1989). "Global Climatic Change." in Scientific American 260, pp. 18-26.

19. BRYAN, K. (1986). "Man's great geophysical experiment: can we model the consequences?" in Oceanus, 29, pp. 36-42.

20. TYNDALL, J. (1863). "On radiation through the Earth's atmosphere." in Phil. Mag., 4, 200.

21. CALLENDAR, G.S. (1938). "The artificial production of carbon dioxide and its influence on temperature." in Q.J.Roy.Meteorol.Soc., 64, 223.

22. NATIONAL ACADEMY OF SCIENCES. (1983). "Changing Climate." Report of the Carbon Dioxide Assessment Committee, National Academy Press, Washington D.C., 360p.

23. SCHLESINGER, M.E., W.L. GATES and Y.J.HAN. (1985). "The role of the ocean in carbon-dioxide-induced climate change: Preliminary results from the OSU coupled atmosphere-ocean general circulation model." In: J.C.J. NIHOUL (Ed.,), Coupled Ocean-Atmosphere Models, Elsevier, 767p.

24. TRABALKA, J.R. and D.E. REICHLE. (1986). The Changing Carbon Cycle: A Global Analysis, Springer Verlag, New York, NY, 592p.

25. FERGUSON, H.L.(Ed.,) (1988). World Conference on The Changing Atmosphere: Implications for Global Security, Toronto, Ontario, June 1988.

26. JONES, P.D., T.M.L. WIGLEY, C.K. FOLLAND, D.E. PARKER, J.K. ANGELL, S. LEBEDEFF and J.E. HANSEN. (1988). "Evidence for global warming in the past decade." in Nature, 332, 790.

27. HARE, F.K. (1988). "Jumping the greenhouse gun?" in Nature, 334, 646.

28. GROVE, J.M. (1988). The Little Ice Age. Methuen. 498p.

29. TAKAHASHI, T., W.S. BROECKER, A.E. BAINBRIDGE and R.F. WEISS. (1980). "Carbonate chemistry of the surface waters of the world oceans." In: E.GOLDBERG, Y. HORIBE and K. SARUHASHI (Eds.,) in Isotope Marine Chemistry, Ochida Rokakuho, Tokyo, 147-182.

30. BACASTOW, R.B. (1976). "Modulation of atmospheric carbon dioxide by the southern oscillation." in Nature, 261, 116-

31. BACASTOW, R.B. (1977). "Influence of the southern oscillation on atmospheric carbon dioxide." In: N.R. ANDERSON and A. MALAHOFF (Eds.,) The Fate of Fossil Fuel CO_2 in the Oceans, Plenum Press, New York, pp. 33-43.

32. KEELING, C.D. and R. REVELLE. (1985). "Effects of El Nino / Southern Oscillation on the atmospheric content of carbon dioxide." in Meteoritics, 20, pp. 437-450.

33. BREWER, P.G. (1986). "What controls the variability of carbon dioxide in the surface ocean? A plea for complete information," in BURTON, J.D., P.G. Brewer and R. Chesselet (Eds.,) (1986). Dynamic Processes in the Chemistry of the Upper Ocean. NATO Conference Series, IV Marine Sciences, Vol. 17, Plenum Press, New York, pp. 215-231.

34. HANSEN, J., A. LACIS, D. RIND, G. RUSSELL, P. STONE, I. FUNG, R. RUEDY and J. LERNER. (1984). "Climate sensitivity: analysis of feedback mechanisms," in J.E. HANSEN and T. TAKAHASHI (Eds.,) Climate Processes and Climate Sensitivity. Maurice Ewing Series, Vol. 5, Am. Geophys. Union, Washington D.C., pp. 130-163.

35. WASHINGTON, W.M., and G.A. MEEHL. (1984. "Seasonal cycle experiment on the climate sensitivity due to a doubling of CO_2 with an atmospheric general circulation model coupled to a simple mixed-layer ocean model." in J. Geophys. res., 89, pp. 9475-9503.

36. WETHERALD, R.T., and S. MANABE. (1986). "An investigation of cloud cover change in response to thermal forcing." in Climatic Change 8, 5-23.

37. WETHERALD, R.T. and S.MANABE. (1988). "Cloud feedback processes in a general cirulation model." in J. Atmos. Sci., 45 (in press).

38. SCHLESINGER, M. (1986). "Equilibrium and transient climatic warming induced by increased atmospheric CO_2." in Climate Dynamics, 1, pp. 35-51.

39. WILSON, C.A. and J.F.B. MITCHELL. (1987). "A doubled CO_2 climate sensitivity experiment with a global climate model including a simple ocean." in J. geophys. res., 92, 13, pp. 315-343.

40. SCHLESINGER, M. & J.F.B. MITCHELL. (1987). "Climate model simulations of the equilibrium climatic response to increased carbondioxide." in Rev. of Geophys., 25, pp. 760-798.

41. SCHLESINGER, M. & Z.C. ZHAO. (1988). "Seasonal climatic changes induced by doubled CO_2 as simulated by the OSU GCM / mixed layer oceanmodel." Report No. 70, Climatic Research Institute, Oregon State University, Corvallis, 73pp.

42. SCHLESINGER, M. (1988). "Model projections of the climatic change induced by increased atmospheric CO_2." Paper presented at Symposium, Louvain la Neuve, August 1988.

43. BROECKER, W.S., & T.H. PENG. (1982). Tracers in the Sea. Eldigio Press, Columbia University, Palisades, New York, 690p.

44. O' BRIEN, J.J. (1986). "An important scientific controversy; oceanic CO_2 fluxes." in J. Geophys. Res., 91 (10) pp. 515-535.

45. DUGDALE, R.C., and J.J. GOERING. (1967). "Uptake of new and regenerated forms of nitrogen in primary productivity." in Limnol. Ocean., 12, pp. 196-206.

46. EPPLEY, R.W., and B.J. PETERSON. (1979). "Particulate organic matter flux and planktonic new production in the deep ocean." in Nature, 282, pp. 677-678.

47. EPPLEY, R.W. (1989). "New production: History, methods, problems." In: BERGER, W.H., V.S. SMETACEK and G. WEFER. (Eds.,) (1988). Productivity of the Ocean: Past and Present. Dahlem Konferenzen, April 1988. John Wiley and Sons Ltd., (Chichester); in press.

48. HAMILTON, J.M., M.R. LEWIS and B.R. RUDDICK. (1989). "Vertical Fluxes of nitrate associated with salt fingers in the world oceans." in J. of Geophys. Res. 94(C2), pp. 2137-2145.

49. DE BAAR, H.J.W., H.M. VAN AKEN, H.G. FRANSZ, G.M. GANSEN, W.W.C. GIESKES, W.G. MOOK and J.H. STEL. (1988). "Towards a Joint Global Ocean Flux Study: Rationale and Objectives." in Oceanography 1988, Proceedings of the Joint Oceanographic Assembly, in press.

50. STUIVER, M.P., P.D. QUAY and H.G. OSTLUND. (1982). "Abyssal water carbon-14 distribution and the age of the world oceans." in Science, 219, pp. 849-851.

51. KEELING, C.D., A.F. CARTER and W.G. MOOK. (1984). "Seasonal, Latitudinal and Secular variations in the Abundance and Isotopic Ratios of Atmospheric CO_2: Results from Oceanographic Cruises in the Tropical Pacific Ocean." in J. Geophys. Res., 89, pp. 4615-4628.

52. McCAVE, I.N. (1975). "Vertical flux of particulates in the ocean." in Deep-Sea Res., 22, pp. 491-502.

53. HONJO, S. (1980). "Material fluxes and modes of sedimentation in the mesopelagic and bathypelagic zones." in J. Mar. Res., 38, pp. 53-97.

54. WAKEHAM, S.G., J.W. FARRINGTON, R.B. GAGOSIAN, C. LEE, H. DE BAAR, G.E. NIGRELLI, B.W. TRIPP, S.O. SMITH and N.M. FREW. (1980). "Organic matter fluxes from sediment traps in the equatorial Atlantic Ocean." in Nature, 286, pp. 798-800.

55. FOWLER, S.W. and G.A. KNAUER. (1986). "Role of large Particles in the Transport of Elements and Organic Compounds Through the Oceanic Water Column." in Prog.

Oceanog., 16, pp. 147-194.

56. MARTIN, J.H., G.A. KNAUER, D.M. KARL and W.W.BROENKOW. (1987). "VERTEX carbon cycling in the Northeast Pacific." in Deep-Sea Res., 34, pp. 267-285.

57. PILSKALN, C. and S. HONJO. (1987). "The fecal pellet fraction of biogeochemical particle fluxes to the deep sea." in Global Biogeochemical Cycles, 1, pp. 31-48.

58. PACE, M.L., G.A. KNAUER, D.M. KARL and J.H. MARTIN. (1987). "Particulate matter fluxes in the ocean: A predictive model." in Nature, 325, pp. 803-804.

59. DE BAAR, H.J.W., FARRINGTON, J.W. and S.G. WAKEHAM. (1983). "Vertical flux of fatty acids in the North Atlantic Ocean." in J. Mar. Res., 41, pp. 19-41.

60. LEE, C. & C. CRONIN. (1984). "Particulate amino acids in the sea: Effects of primary productivity and biological decomposition." in J. Mar. Res., 42, pp. 1075-1097.

61. WAKEHAM, S.G., C. LEE, J.W. FARRINGTON and R.B. GAGOSIAN. (1984). "Biogeochemistry of particulate organi matter in the oceans: results from sediment trap experiments." in Deep-Sea Res., 31, pp. 509-528.

62. WATSON, A.J. and M. WHITFIELD. (1985). "Composition of particles in the global ocean." in Deep-Sea Res., 32, pp. 1023-1039.

63. DEUSER, W.G. (1986). "Seasonal and interannual variations in deep-water particle fluxes in the Sargasso Sea and their relation to surface hydrography." in Deep-Sea Res., 33A, pp. 225-246.

64. SUGIMURA, Y. and Y. SUZUKI. (1988). "A high temperature catalytic oxidation method for the determination of non-volatile dissolved organic carbon in seawater by direct injection of a liquid sample." in Mar. Chem., 24, pp. 105-131.

65. KARL, D.M., G.A. KNAUER and J.H. MARTIN. (1988). "Downward flux of particulate organic matter in the ocean: a particle decomposition paradox." in Nature, 332, pp. 438-441.

66. CHO, B.C.and F. AZAM. (1988). "Major role of bacteria in biogeochemical fluxes in the ocean's interior." in Nature, 332, pp. 441-443.

67. WILLIAMS, P.M. and E.R.M. DRUFFEL. (1988). "Dissolved Organic Matter in the Ocean: Comments on a Controversy." in Oceanography, 1, pp. 14-17.

68. TOGGWEILER, J.R. (1989). "Is the downward dissolved organic matter (DOM) flux important in carbon transport?" in W.H. BERGER, V.S. SMETACEK and G. WEFER (Eds.,). (1988). Productivity of the Ocean: Past and Present. Dahlem Konferenzen, April 1988. John Wiley and Sons Ltd (Chichester); in press.

69. WILLIAMS, P.J. leB. (1981). "Incorporation of microheterotrophic processes into the classical paradigm of

the planktonic food web." in <u>Kieler Meeresforsch.</u>, 5, pp. 1-28.

70. WILLIAMS, P.J. LeB. (1984). "Bacterial production in the marine food chain: The emperor's new suit of clothes?" in M.J.R. FASHAM (Ed.,) <u>Flows of Energy and Materials in Marine Ecosystems</u>, Plenum Press, New York, pp. 271-299.

71. AZAM, F., T. FENCHEL, J.G. FIELD, J.S. GRAY, L.A. MEYER-REIL and F. THINGSTAD. (1983). "The ecological role of water-column microbes in the sea." in <u>Mar. Ecol. Prog. Ser.</u>, 10, pp. 257-263.

72. HOBBIE, J.E. and P.J. LeB. WILLIAMS. (Eds.,). (1984). "Heterotrophic activity in the sea." NATO conference Series IV: Marine Science, Plenum Press, Vol. 15, xv + 569 pp.

73. BIRD, D. and J. KALFF. (1984). "Empirical relationship between bacterial abundance and chlorophyll concentration in fresh and marine waters." in <u>Can. J. Fish. Aquat. Sci.</u>, 41, pp. 1015-1023.

74. HODSON, R.E. and F. AZAM. (1977). "Size distribution and activity of matine microheterotrophs." in <u>Limnol. Oceanogr.</u> 22, pp. 492-501.

75. DUCKLOW, H.W., D.A. PURDIE, P.J. LeB. WILLIAMS and J.M. DAVIES. (1986). "Bacterioplankton: A sink for carbon in a coastal plankton community." in <u>Science</u>, 232, 865-867.

76. PLATT T. (1985). "Structure of the marine ecosystem: Its allometric basis." in R.E. ULANOWICZ and T. PLATT (Eds.,). Ecosystem Theory for Biological Oceanography. <u>Can. Bull. Fish. Aquatic. Sci.</u>, 213, pp. 55-64.

77. FROST, B.W. (1984). "Utilization of phytoplankton production in the surface layer." in Global Ocean Flux Study: Proceedings of a Workshop, National Academy Press, Washington D.C., pp. 125-135.

78. FASHAM, M.J.R. (1985). "Flow analysis of materials in the marine euphotic zone." in <u>Can. Bull. Fish. Aquat. Sci.</u>, 213, pp. 139-162.

79. FRANSZ, H.G. and J.H.G. VERHAGEN. (1985). "Modelling research on the production cycle of phytoplankton in the Southern Bight of the North sea in relation to riverborne nutrient loads." in <u>Neth.J.Sea Res.</u>, 19, pp. 241-250.

80. DEUSER, W.G., E.H. ROSS, C. HEMLEBEN and M.SPINDLER. (1981). "Seasonal change in species composition, numbers, mass, size and isotopic composition of planktonic foraminifera settling into the deep Sargasso Sea." in <u>Paleogeography, Paleoclimatology and Paleoecology</u>, 33, pp. 103-127.

81. VOLK, T. and M.I. HOFFERT. (1985). "Ocean Carbon Pumps: Analysis of Relative Strengths and Efficiencies in Ocean-Driven Atmospheric CO_2 changes." in E.T.

SUNDQUIST and W.S. BROECKER (Eds.,). The Carbon Cycle and Atmospheric CO_2: Natural Variations Archean to Present. Geophys. Monograph Series, Vol. 32, Am. Geophys. Union, Washington D.C., pp. 99-110

82. SKIRROW, G. (1975). "The dissolved gases - Carbon Dioxide." in J.P. Riley and G. Skirrow (Eds.,) Chemical Oceanography, New York, Academic Press, pp. 1-192.

83. KENNETT, J.P. (1982). Marine Geology, Prentice Hall, Inc. Englewood Cliffs, N.J. 813 pp.

84. BAES, C.F. and G.G. KILLOUGH. (1986). "Chemical and biological processes in CO_2-Ocean Models." in J.R. Trabalka and D.E. Reichle (Eds.,). The Changing Carbon Cycle : A Global Analysis, New York, Springer Verlag, 329-347 pp.

85. SHAFFER, G. (1989). "A model of biogeochemical cycling of phosphorus, nitrogen, oxygen and sulfur in the Ocean: one step toward a global climate model." in J. of Geophys. Res. 94(C2), pp. 1979-2004.

86. PLASS, G.N. (1972). "Relationship betweem atmospheric carbon dioxide amount and the properties of the sea." in Environ. Sci. Technol. 6(8), pp. 736-740.

87. HOFFERT, M.I. (1974). "Global distribution of atmospheric carbon dioxide in the fossil fuel era: a projection." in Atmospheric Environment 8, pp. 1225-1249.

88. SUNDQUIST, E.T. (1988). "Implications of Pleistocene CO_2 changes for the long term buffering of antropogenic CO_2." in EOS 69(44), 1236.

89. BERNER, R.A. and A.C. LASAGA. (1989). "Modeling the geochemical Carbon Cycle." in Scientific American March 1989, pp. 54-61.

90. SIEGENTHALER, U. and H. OESCHGER. (1978). "Predicting future atmospheric carbon dioxide levels." in Science 199(4237), pp. 388-395.

91. TRABALKA, J.R., J.A. EDMONDS, J.M. REILLY, R.H. GARDNER and D.E. REICHLE. (1986). "Atmospheric CO_2 projections with globally averaged carbon cycle models." in J.R. Trabalka and D.E. Reichle (Eds.,). The Changing Carbon Cycle : A Global Analysis, New York, Springer Verlag, pp. 534-560.

92. MARLAND, G. (1986). "Technical fixes for limiting the increase of atmospheric CO_2: A review." Institute for Energy Analysis, Oak Ridge Associated Universities, Oak Ridge, Tennesee, U.S.A., August, Unpublished manuscript.

93. DYSON, F.J. (1976). "Can we control the amount of carbon dioxide in the atmosphere?" IEA(0)-76-4, Institute for Energy Analysis, Oak Ridge Associated Universities, Oak Ridge, Tennessee, July.

94. DYSON, F.J. and G. MARLAND (1979). "Technical fixes for the climatic effects of CO_2." in W.P. Elliott and

L. Machta, Workshop on the global effects of Carbon dioxide from fossil fuels, Miami Beach, Florida, March 7-11, 1977, U.S. Department of Energy, CONF-770305, pp. 111-118.

95. BROECKER, W.S. (1977). Unpublished manuscript as cited by Marland (1986)

96. MARCHETTI, C. (1975). "On geoengineering and the CO_2 problem," International Institute for Applied Systems analysis, Laxenburg, Austria, July.

97. MARCHETTI, C. (1978). "Constructive solutions to the CO_2 problem," International Institute for Applied Systems analysis, Laxenburg, Austria.

98. WHITEHEAD, J.A. (1989). "Giant Ocean Cataracts." in Scientific American February 1989, pp. 36-43.

99. BAES, C.F., S.E. BEALL and G. MARLAND. (1979). "Options for collection and disposal of carbon dioxide from concentrated sources." in U.S. Department of Energy Environmental Control Symposium, Proceedings, DOE/EV-0046, Vol. 1, pp. 260-271.

100. BAES, C.F., S.E. BEALL, D.W. LEE and G. MARLAND. (1980a). "Options for the collection and disposal of carbon dioxide." ORNL-5657, Oak Ridge National Laboratory, Oak Ridge, Tennessee.

101. BAES, C.F., S.E. BEALL, D.W. LEE and G. MARLAND. (1980b). "The collection, disposal and storage of carbon dioxide." in W. Bach, J.Pankrath and J. Williams (Eds.) Interactions of Energy and Climate, pp. 495-519, Reidel Publishing Co., Boston, Massachusetts.

102. HOHMANN, R.P. and A.H. KWAI. (1980). "A survey of methods for isolating and containing gaseous carbon dioxide as nonvolatile products." ORNL/MIT-291, Oak Ridge National Laboratory, Oak Ridge, Tennessee.

103. HOFFERT, M.I., Y-C. WEY, A.J. CALLEGARI and W.S. BROECKER. (1979). "Atmospheric response to deep-sea injections of fossil fuel carbon dioxide." in Climatic Change 2, pp. 53-68.

104. LYSEN, E.H. (1989). "The absorption of carbon dioxide by the Oceans, Amersfoort, April 1989 (in Dutch). Dutch Ministry of the Environment (VROM/DGM) ELMI - Project.

105. BARON, S., and M. STEINBERG. (1975). "The economics of the production of liquid fuel and fertilizer by the fixation of atmospheric carbon and nitrogen using nuclear power." BNL 20273, Brookhaven National Laboratory, Upton, Long Island, New York.

106. ALBANESE, A.S., and M. STEINBERG. (1980). "Environmental control technology for atmospheric carbon dioxide." Final report, DOE/EV- 0079, U.S. Department of Energy.

107. HORN, F.L., and M. STEINBERG. (1982). "Posible storage sites for disposal and environmental control of atmos-

pheric carbon dioxide." BNL-51597, Brookhaven National Laboratory, Upton, Long Island, New York.

108. STEINBERG, M. (1983). "An analysis of concepts for controlling atmospheric carbon dioxide." DOE/CH/00016-1, U.S. Department of Energy.

109. STEINBERG, M., H.C. CHENG and F. HORN. (1984). "A systems study for the removal, recovery and disposal of carbon dioxide from fossil fuel power plants in the U.S." DOE/CH/00016-2, U.S. Department of Energy.

110. STEINBERG, M., and H.C. CHENG. (1988). "A systems study for the removal, recovery and disposal of carbon dioxide from fossil fuel power plants in the U.S." Brookhaven National Laboratory, New York, May 1984. For: U.S. Department of Energy, Office of Energy Research, Office of Basic Energy Sciences, carbon dioxide research Division.

111. KAPLAN, L.J. (1982). "Cost saving process recovers CO_2 from power plant Fuel Gas." in Chem. Eng. November 29, p. 30.

112. ELLINGTON, R.T., L. WARZEL, B. ACHILLADELIS, K. SALDANHA and M.J. MUELLER. (1984). "Scrubbing CO_2 from plant exhausts provides economic sources of gas for EOR Projects." in Oil and Gas Journal, October 15, pp. 112-124.

113. BLOK, K., C. HENDRIKS and W. TURKENBURG. (1989). "The role of carbon dioxide removal in the reduction of the greenhouse effect." Contribution to the IEA/OECD Expert Seminar on Energy Technologies for Reducing Emissions of Greenhouse gases, Paris, 12-14th April, 1989.

114. OESCHGER, H. and B. STAUFER. (1986). "Review of the history of atmospheric CO_2 recorded in Ice Cores," in J.R. Trabalka (Ed.,) Atmospheric Carbon Dioxide and the Global Carbon Cycle, Oak Ridge National Laboratory, Oak Ridge, U.S. DOE/ER-0239, pp. 89-108.

115. ROTTY, R.M. and C.D. MASTERS. (1985). Carbon dioxide from fossil fuel combustion: Trends, resources and technological implications," in J.R. Trabalka (Ed.,) Atmospheric Carbon Dioxide and the Global Carbon Cycle, Oak Ridge National Laboratory, Oak Ridge, U.S. DOE/ER-0239, pp. 63-80

116. MARLAND, G. (personal communication).

117. PERRY, A.M. (1986). "Possible changes in the future Use of fossil fuel to limit environmental effects." in J.R. Trabalka (Ed.,) Atmospheric Carbon Dioxide and the Global Carbon Cycle, Oak Ridge National Laboratory, Oak Ridge, U.S. DOE/ER-0239, pp. 561-570.

118. CHENG, H.C., M. STEINBERG and M. BELLER. (1986). "Effects of energy technology on global CO_2 emissions," DOE/NBB-0076, pp. 1-92.

119. ROTTY, R.M. and G. MARLAND. (1986). "Fossil fuel combustion: recent amounts, patterns and trends of CO_2." in J.R. Trabalka and D.E. Reichle (Eds.,). The Changing Carbon Cycle - A Global Analysis, New York, Springer Verlag, pp. 474-490.
120. BARNOLA, J.M., D. RAYNAUD, Y.S. KORETKEVICH and C. LORIUS. (1987). "Vostok ice core provides 160,000-year record of atmospheric CO_2." in Nature, 329, pp. 408-414.

ACKNOWLEDGEMENTS

Upon invitation by the Ministry of Housing, Physical Planning and the Environment (VROM) this paper was written as a contribution to the Symposium 'Energy and Climate' on 27 September, 1989 at Utrecht, The Netherlands. Communications with Gregg Marland (Carbon Dioxide Center, Oak Ridge National Laboratory, USA), Said Zwerver (Federal Institute for Public Health and Environment, RIVM) and Erik Lysen (Consulting Engineer) have proven to be pleasant and fruitful. We are indebted to Henk Postma, Pieternel Montijn and Mario Hoppema for proofreading the draft manuscript. Support of Michel Stoll by the CO_2 Working Group of NWO is gratefully acknowledged.

DISPOSAL OF CARBON DIOXIDE IN DEPLETED NATURAL GAS RESERVOIRS

A.C. van der Harst, A.J.F.M. van Nieuwland
Nederlandse Aardolie Maatschappij B.V.
P.O. Box 28 000
9400 HH   ASSEN
The Netherlands

ABSTRACT. The emission of carbon dioxide into the atmosphere, which is one of the causes of the greenhouse effect, could be reduced by the removal of carbon dioxide from stack gases of power plants and subsequent injection of the removed carbon dioxide in depleted gas reservoirs. In The Netherlands there are some 220 gas reservoirs of which 90 are in production. The largest field, Groningen, with initial gas reserves of some 2,500 mrd m^3 has a potential for carbon dioxide storage of 8×10^9 ton. The field cannot play an immediate role in the combat against the greenhouse effect, since its presently estimated depletion date is around the middle of the next century. Other Dutch onshore fields have a storage potential of 1.3×10^9 ton of carbon dioxide divided over about 100 reservoirs. These fields will gradually become available starting from about 2000 onwards. The cost of transport and injection of carbon dioxide in onshore reservoirs is estimated at Dfl 7,50/ton of carbon dioxide.

1. INTRODUCTION

Carbon dioxide emissions into the atmosphere can be reduced, inter alia, by improving energy efficiency, reforestation, and renewable energy. The option to remove carbon dioxide from stack gases of power plants is discussed in references 1 and 4. Options to dispose the removed carbon dioxide are injection into oceans, chemical capture or injection into depleted natural gas reservoirs. In The Netherlands there are, besides the giant Groningen field, a large number of smaller reservoirs, which could be used for this purpose. In this article the technical feasibility, availability in time, storage potential and cost of the injection option are discussed for comparison with other carbon dioxide disposal options.

2. PRODUCTION FROM GASFIELDS IN THE NETHERLANDS

2.1 Typical gasfield depletion

Most non associated natural gas reservoirs in The Netherlands are depletion type reservoirs. In figure 1 a
scheme of a typical gas field depletion process is given.
In the period after start-up the production rate is kept
constant by limiting the well production. During the production life the tubing head pressure of the well drops
constantly as a result of declining reservoir pressures and
below a certain critical tubing head pressure, compression
facilities will be needed to maintain the required production rate while the tubing head pressure declines further.
Second stage compression can be installed later if economically justified.

Abandonment of a field involves the removal of all
surface facilities and the plugging of the wells, and takes
place at the end of the field's life. The time of abandonment is primarily an economic decision. For continuation of
operation of a field, the ongoing development and operating
cost need to meet the economic criteria of the relevant
owning oil/gas company. Under today's tax regimes and gas
prices, abandonment is assumed to take place at reservoir
pressures between 20 and 50 bar. The reservoir pressure at
which abandonment takes place tends to be mainly reservoir
size and depth related.

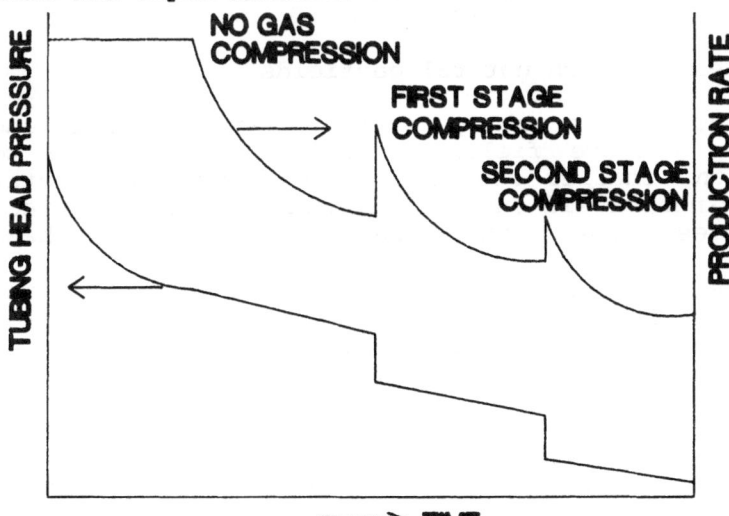

Figure 1. Scheme of typical gasfield depletion

2.2   Dutch gas production policy

Since in 1963 production of the Groningen field star-ted, a large number of smaller reservoirs, both onshore and offshore, were discovered (see figure 2). Of the 220 reser-voirs outside Groningen some 90 are in production.

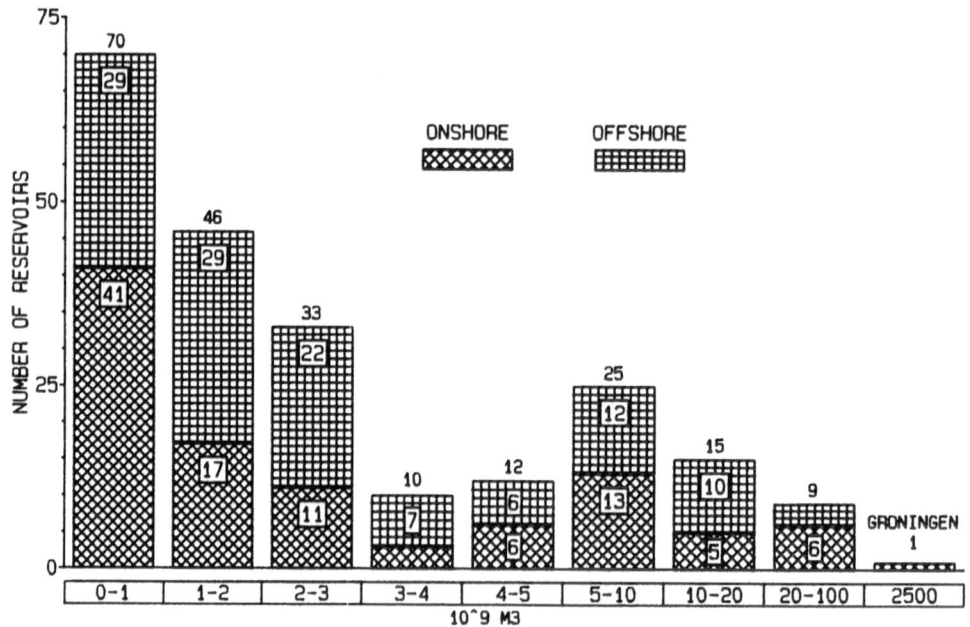

Figure 2. Sizes of Dutch natural gasfields

Present government policy is to preserve Groningen and preferentially produce fields outside Groningen. Due to this policy the smaller fields are being depleted faster than the Groningen field and in a number of cases compres-sion facilities have already been installed in these fields.

Abandonment of the first fields will take place in the late-1990s. Abandonment of Groningen is presently estimated to take place around 2050. First stage compression in the Groningen field is now estimated to be installed from the mid 1990s.

If disposal of carbon dioxide in reservoirs is a rea-listic option, the best point in time to start injection of carbon dioxide into the reservoir is at the abandonment stage. Much of the existing facilities, especially wells, pipelines and plot space, can then still be used.

3. CARBON DIOXIDE STORAGE POTENTIAL

3.1 Carbon dioxide characteristics

The storage potential for carbon dioxide in depleted reservoirs depends on the following aspects:
- the density;
- the viscosity;
- the gas deviation factor or z-factor;
- the gas temperature;
- the reservoir pressure;
- the reservoir permeability thickness.
- the well completion;

The density is of course dependent on the pressure. It increases rapidly with higher pressures; at atmospheric pressure the density is about 2.5 times the density of natural gas and at 300 bar the density is 3.5 times as large as the density of natural gas. This means that injection of carbon dioxide will require lower tubing head pressures than injection of natural gas at the same reservoir pressures.

The viscosity is higher than for natural gas and also increases rapidly with pressure. The z-factor deviates much more rapidly from the ideal gas behaviour than natural gas. The subsurface temperatures vary from 20 degrees C just below surface to approximately 100 degrees C at a reservoir depth of 3,000 m.

It is assumed that gas reservoirs will be abandoned at pressures between 20 and 50 bar. During refilling with carbon dioxide the reservoir pressure can be allowed to increase to the initial pressure, prior to natural gas production. For Rotliegendes reservoirs as found in the north of The Netherlands the initial pressures range from 280 bar at 2,500 m to 350 bar at 3,000 m depth. It is not recommended to repressurise significantly above the initial reservoir pressure as the integrity of the sealing cap-rock cannot be guaranteed at higher pressures.

The permeability thickness is a quantity used in well inflow performance description. It is the product of the permeability and the thickness of the reservoir. The higher the value of this product the better the injection performance will be. In this article the reservoir properties of a typical Rotliegendes reservoir have been used as most gas fields in the North of The Netherlands are completed on this formation where permeability ranges from 100-300 m Darcy and have thicknesses between 70 and 200 m.

Most gaswells in The Netherlands are equipped with 5" or 5½" tubing. At flowrates of 3 mln m^3/d and a reservoir pressure of 300 bar the required compressor discharge pressure for injection is about 210 bar. The most recently drilled wells are often completed with 7" tubing which improves productivity and thus the possible future injection performance considerably.

3.2 Available reservoir volumes in the netherlands

In The Netherlands, gas fields are found at depths ranging from approximately 2,000 to 3,000 m. The initial pressures lie typically in the order of 250 to 350 bar. At 300 bar the carbon dioxide density is about 650 kg/m^3 or 3.5 times as large as the natural gas density. One ton of carbon dioxide will, at reservoir conditions, occupy a volume of 1.5 m^3 which is equivalent to 325 m^3 of natural gas at standard conditions (1 bar pressure and 15 C).
So each 10^9 m^3 of gas in the reservoir can be replaced by approximately 3.1 x 10^6 ton of carbon dioxide if it is disposed at a reservoir pressure of 300 bar.

The total recoverable gas reserves in The Netherlands are estimated at about 3,400 mrd m^3 of which 1,500 mrd m^3 have been produced until now. The division over Groningen, onshore and offshore fields is shown in table 1 (Reference 3).

TABLE 1. Division of reserves over groningen, onshore and offshore fields (mrd m^3)

Reservoir	Initial Reserves	Produced 01.01.89	Remaining Reserves
Groningen	2,508	1,184	1,324
Onshore fields	429	196	233
Offshore fields	477	157	320
Total	3,414	1,537	1,877

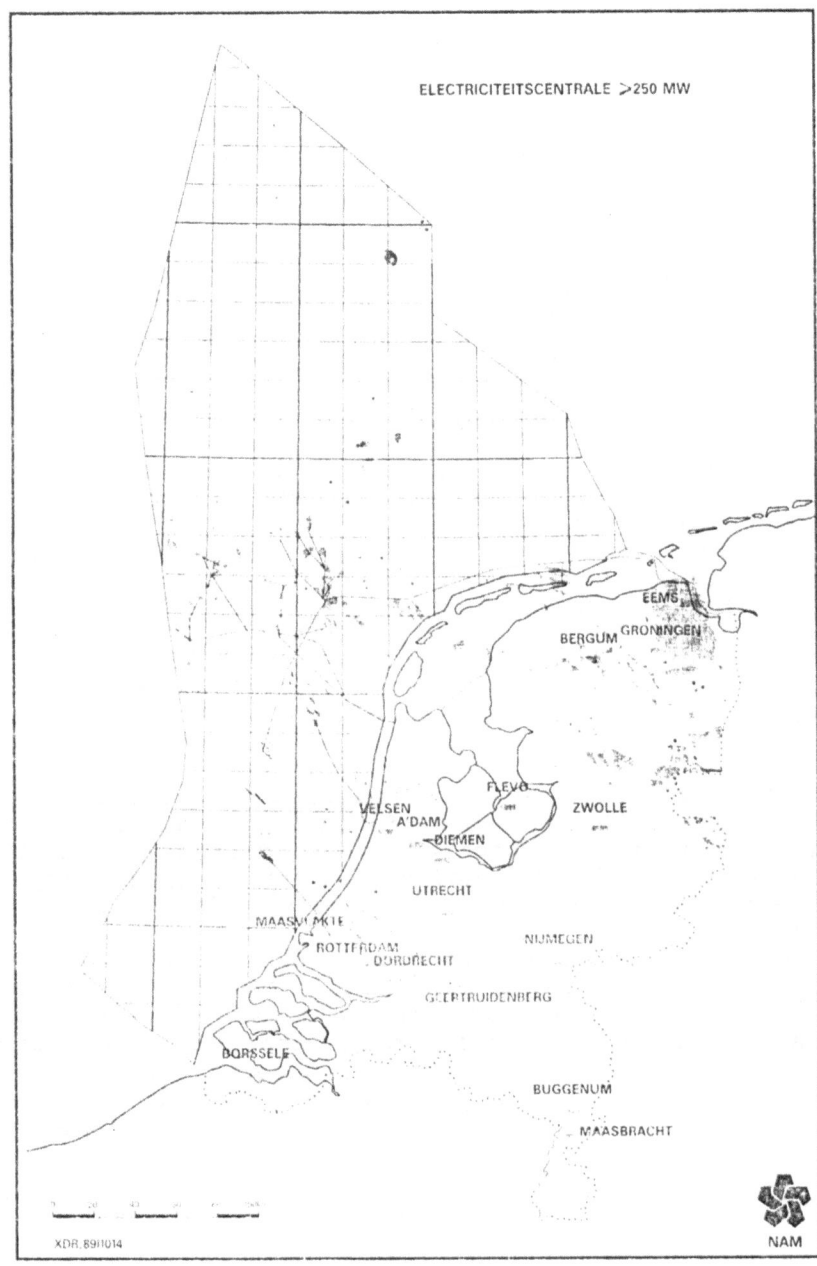

Figure 3. Location of gas reservoirs and electric power
plants in The Netherlands

3.2.1 **Groningen.** As discussed earlier Groningen will be available at the middle of the next century. In view of this late abandonment date, it cannot play an immediate role to combat carbon dioxide emissions into the atmosphere. Due to its gigantic size about 8 mrd ton of carbon dioxide could be stored in the Groningen reservoir. Based on a carbon dioxide emission from a 600 MW power plant of 2.8 mln ton/year, it would take 50 power plants 55 years to refill the reservoir.

3.2.2 **Onshore fields.** Onshore fields are mainly located in the northern part of The Netherlands around Groningen (Figure 3). There are also a number of fields in production in the province of North Holland.

Total carbon dioxide storage potential is about 1.3 mrd ton of carbon dioxide. The largest field Annerveen has a volume of 60 mrd m³ of recoverable reserves. It would take one 600 MW coal fired power plant 66 years to fill up this reservoir. The injection rate of 450 tons/hour would require three to four injection wells.

The availibility of the Annerveen reservoir for potential carbon dioxide disposal will depend on the results of ongoing studies related to gas storage and peak shaving potential.

Other examples of relatively large fields that will be abandoned around or before 2000 are Tietjerk (15 mrd m³), Bergermeer (17 mrd m³), Norg (6 mrd m³) and Grootegast (3.6 mrd m³).

3.2.3 **Offshore fields.** As shown in Figure 3 offshore reservoirs are evacuated to shore via common submarine pipelines, thus reducing the high costs associated with development of offshore fields. New fields are being developed while others are being depleted to optimally load these submarine lines.

The costs for disposal of carbon dioxide in offshore fields will therefore be considerably higher than for onshore fields, since almost none of the existing field facilities can be reused. Onshore compression facilities with high pressure (250 bar) pipelines connected to existing well platforms will probably be the only technical/operational feasible alternative. Compared to onshore injection, an offshore development is not very attractive. The total offshore capacity volume of 477 mrd m³ is therefore not considered for carbon dioxide disposal.

3.3 World carbon dioxide storage potential

Current world proven reserves of natural gas are estimated to be more than 100,000 milliard m³, which would satisfy present demand for about 55 years. Approximately

80% of these reserves are contained in just 10 countries. Nearly 40% are in the USSR, around half of which are in three supergiant fields - Urengoy, Yamburg and Bovanenkovskoye - in West Siberia. A further 30% are located in the Middle East of which Iran has a 40% share. The U.S.A. and Western Europe account for 11% of total proven gas reserves. Together with the gas produced prior to this date the storage potential is estimated at 150,000 mrd m³ which is about 60 times the potential of the Groningen field.

4. ULTIMATE RECOVERY ASPECTS

Inert gas injection at the last stage of gas field life is a possibility to increase ultimate hydrocarbon recovery. Based on energy prices and cost to date this technique could only be considered for relatively large gasfields. In case carbon dioxide injection is considered for environmental purposes all gas fields could benefit from an increase in ultimate recovery.

Injection of carbon dioxide into oilfields is a wellknown technique for improving the recovery, mainly for light oil (20 to 30 deg API). The disadvantage of this technique from a disposal viewpoint is that the carbon dioxide is rather quickly produced back to the surface with the oil, where it has to be separated from any other produced gases in order to be reinjected, which makes it a rather expensive carbon dioxide disposal process.

5. INJECTION INTO AQUIFERS

Aquifers are an abundantly available disposal option. They have been used by several companies for underground gas storage for peak shaving purposes. The major disadvantages of aquifers are the absence of any infrastructure and the uncertainty about the sealing capacity of the cap rock. Wells, pipelines and compressors will have to be installed especially for the carbon dioxide disposal project. This makes aquifer disposal very expensive and unattractive compared to fully depleted gas fields.

6. CARBON DIOXIDE DISPOSAL FACILITIES

6.1 Process description

A schematic diagram for the disposal of carbon dioxide is shown in figure 4. After removal from the flue gas of the power plant, the carbon dioxide needs to be compressed for pipeline transport. In view of the corrosivity of car-

186

bon dioxide in combination with free water, dehydration is
required to a water dewpoint of -2 C at 65 bar to avoid
possible condensation of water in the carbon steel pipeline
at all times.

The transport pipeline will connect the power plant
site and the reservoir site. The carbon dioxide will be
further compressed at the reservoir site to pressures
around 210 bar and distributed via high pressure pipelines
to the injection wells.

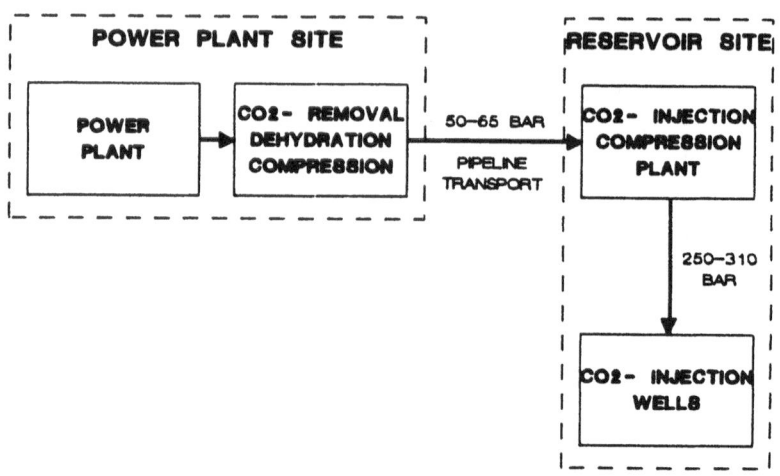

Figure 4. Schematic diagram for carbon dioxide disposal

6.2 Cost estimate of carbon dioxide disposal facilities

The costs for carbon dioxide removal and dehydration
are discussed in Reference 4. The pipeline costs will depend on the distance between power plants and reservoir
site. As can be seen from Figure 4 onshore gas reservoirs
are all located in the Northern part of The Netherlands and
the existing power plants are built evenly over the country. For new power plants the location can of course be
optimized. On average it is assumed that carbon dioxide
will be transported over a distance of 100 kilometres. For
a coal fired 600 MW power plant with a carbon dioxide production of 450 ton carbon dioxide/hr and 6300 hrs/year in
service, the costs are estimated at Dfl. 60 million.

Total investments are:		mln Dfl
- transportation pipeline		60
- compression plant		50
- wells (reuse of existing ones)		10
	Total	120

Yearly operating costs are: mln Dfl

- interest (7%) and depreciation (25 years)   10.2
- operation and maintenance   3.0
- electricity cost   7.6
                                               ────
                                       Total   20.8

This corresponds to a cost of Dfl 7.50 per ton of carbon dioxide. The cost for removal and dehydration of carbon dioxide are almost a factor 10 higher.

7. RISKS

7.1 Formation weakening

Together with water in the reservoir formation carbon dioxide will form an acid solution. This could weaken the strength of the formation. Research would be required to find out if this is a real problem.

7.2 Leaking of carbon dioxide to the surface

The possibility of carbon dioxide, once injected, to leak back to the surface is very small if depleted gas reservoirs are used. The integrity of the cap rock over geological times has been proven by the fact that the gas remained in place until it was discovered. Therefore the reservoir should not be exposed to pressures above its original pressure. A more likely route for leaks would be via the wells that penetrate the reservoir. The casings have been cemented and at abandonment cement plugs will be used to isolate the gasbearing formation from the layers above. There is no evidence so far that the cement would deteriorate in the long run and even if it would the leakage rates would be very small and the carbon dioxide would probably be absorbed by the waterbearing formations above the reservoir.

7.3 Safety measures

Carbon dioxide is dangerous at high concentrations when it displaces air. Leak detection equipment will be required. In principle however, no more stringent measures are required than applied in gas transport/injection.

8. CONCLUDING REMARKS

Injection of carbon dioxide generated by powerplants into depleted onshore gas reservoirs could be an efficient way of reducing the emission of carbon dioxide into the atmosphere. Utilisation of aquifers and oil reservoir is less attractive.

Injection of carbon dioxide on a large scale will only be possible when the giant gas fields, like Groningen, are fully depleted. These fields will only be available in the second half of the 21st century. Injection into smaller fields on a smaller scale could start soon after 2000.

The possible positive impact of carbon dioxide injection on the ultimate recovery of gasfields will have to be investigated in more detail. Start-up of a carbon dioxide injection project should take place just prior to the complete depletion of the gasfield, possibly extending the period of hydrocarbon gas production by several years.

The estimated cost for carbon dioxide disposal including transport is estimated at Dfl 7.50 per ton of carbon dioxide. (Note that the cost of carbon dioxide removal and dehydration would have to be added to this cost, see Ref. 4).

REFERENCES

1. The rate of carbon dioxide removal in the reduction of the greenhouse effect; by K. Blok, C. Hendriks and W. Turkenburg; paper for the IAE/OECD Greenhouse effect conference in Paris (April 1989).
2. Natural gas and oil in The Netherlands 1988, Department of Economic Affairs, May 1989 (in Dutch).
3. International Petroleum Encyclopedia; Penn Well Publishing G. Tulsa (1988).
4. The recovery of carbon dioxide from power plants; C.A. Hendriks, K. Blok, W.C. Turkenburg, RIVM Symposium "Energy and Climate" September 1989.

AN ALL-ELECTRIC SOCIETY FOR LESS CO2?

Jan H.C. van der Veer
Toine W.M. van Wunnik
N.V. KEMA
P.O. Box 9035
6800 ET Arnhem
The Netherlands

ABSTRACT. The application of more electricity, going into the direction of an all-electric society, can result in a reduction of the CO_2 emissions of 5 to 10 percent for the situation in The Netherlands. For most other countries better figures are to be expected.

1. INTRODUCTION

Our society is becoming more and more electric. The share of electricity in the overall energy package is growing. Developments in both the short and the long term make it believable that this trend will continue. Reflecting on the CO_2 economy, the question arises whether these developments are favourable or not.

As the figures in this paper are originated from the Dutch situation the final use of energy in The Netherlands is related to figures for Western Europe and the world in Table 1.

Table 1 Final use of energy in 1987 (1, 2)

	Total final use of energy		Final use of electricity	
	in PJ	in % of global	in PJ	in % of total
The Netherlands	2,200	1	240	11
Western Europe	44,100	16	5,800	13
Global*	267,600	100	34,800	13

*) 1985 figures

189

The all-electric phenomena is here interpreted in a wide sense. Also the generation of electricity by combined heat and power is included as favourable for energy savings and thus resulting in less CO_2.

This paper presents the results for the CO_2 impact of a further introduction of electricity in existing and new markets related to various ways of power generation.

On the demand side the following markets are analysed in more detail:
- residential areas with all-electric houses, including electricity for space heating
- residential areas with district heating
- traffic with large scale introduction of the electric car
- industry.

On the generation side several options are reviewed:
- the high-efficiency combined-cycle power unit
- the standard coal-fired steam boiler/steam turbine power unit
- the integrated coal gasification combined-cycle power unit
- the combined heat and power unit based on the combined cycle
- renewable energy resources.

For the year 2015, this CO_2 impact has been quantified, based on realizable options. For the time after, until 2050 approximately, qualitative indications are given of the long-term developments.

The results are presented in the form of a matrix, with on one side the markets and on the other side the generating options. Presenting the results this way offers the opportunity to create different combinations by interpolating the figures, so that everyone is able to determine the best solution for his specific situation.

2. ENERGY DEMAND

In an all electric society the use of electric energy will intensify in various sectors of the community. The impact of intensifying the use of electricity and a far more efficient use of energy by combined heat and power will be considered for the domestic energy demand including electric space heating, district heating, traffic and industry. The contribution of these sectors to the final use of in the Dutch society in 1987 (2) was respectively 22 percent, 16 percent and 42 percent.

2.1 All-electric housing (3, 4, 5)

At the moment the domestic energy demand in the Netherlands is mainly met using natural gas and electricity.

Annually gas consumption for space heating and hot water in an average insulated house is in the order of 37.5 GJ. This energy demand is equivalent to 1360 cubic meters of natural gas. The electricity demand is nearly 3,000 kWh per annum. Future houses will be insulated in such a way that gas consumption for space heating will decrease to 20 GJ (i.e. 670 cubic meters) or less, thus following the trend of less energy demand. At this point a distribution grid for natural gas as well as for electricity in residential areas becomes too expensive. As electric energy is more flexible the option of electric space heating in combination with hot water supply was studied in order to bypass the neccessity of a gas distribution network. This resulted in the design of an all-electric house.

It is a well-insulated dwelling with a closed air ventilation system. This system is integrated with a central air heating system driven by a boiler with an electrical heat pump providing heat recovery from the ventilation air. The energy demand of an all-electric house will be about 15 GJ for space heating and 7.5 GJ for hot water supply. The assumption here is that in 2015 about 540,000 all-electric houses will have been realised. This number of houses is 25 percent of the houses to be built from now until 2015. This will result in an increase of electric energy demand by 2.3 TWh/year, assuming a coefficient of performance (cop) of the heat pump of 2.5. The increase of the electric energy demand will replace a domestic gas demand of 23 PJ per annum.

2.2 District heating

In order to improve the overall efficiency of our energy supply, it is necessary to use the waste heat of power generation increasingly for district heating (6). At the moment an equivalent of 200,000 houses is heated by district heating. This number now is 4 percent of the total number of houses in the Netherlands. Here the assumption is that in 2015 the heat demand of another 1.5 million houses will be provided by district heating, about 25 percent of the total number of houses then. Furthermore, it is assumed that these houses have moderate insulation, which results in an energy demand for heating of 37.5 GJ/year. This will give a net increase of the heat production of power stations of 148 PJ with natural gas. This figure is the difference between the domestic energy demand, supplied with natural gas with an efficiency between 0.6 and 0.85 (at a high heating value), and the gas supply for power generation with a thermal efficiency of 0.3.

On the other hand electric energy production will increase as well. With an efficiency of 0.50 the increase of electric energy will amount to 30.6 TWh/year. This energy will replace energy production by less efficient or more polluting power stations.

2.3 Traffic

The energy demand in the road traffic sector at the moment is in the order of 280 PJ/year (7). About 20 percent is due to long distance road transport; 77 percent is consumed by private cars and short distance road transport.

A susbstantial penetration of electric energy in the traffic sector can be realised with the introduction of small electric vehicles. The electric car, which has an efficiency of 50 percent, will be used mainly in urban areas due to its limited radius of action. It is assumed that in 2015 20 percent of the total kinetic energy demand in the short distance traffic sector will be met by electric energy. As a result, the net electric energy demand will increase by 5.2 TWh/year. Furthermore, less energy will be needed in the form of automotive fuels. This decrease will be 41.2 PJ, taking into account the increased and better efficiency of modern car engines in the year 2015.

2.4 Industry

The demand for more efficient energy production in industry will increase the penetration of combined heat and power. The assumption is made that in the year 2015 the installed generating capacity will have grown another 1,000 MWe, with an effective use of 6000 hours/year. This will result in a net electric energy production of 6 TWh and a net heat production of 54 PJ. The high efficiency of combined power generation will result in less power and heat generation by less efficient methods.

2.5 Summary on energy demand

Table 2 represents the shift in energy demand between heat and electricity in a more electrical society. The total shift in energy demand is about 13 percent of the final energy demand in 1987 and is mainly due to a different way of power generation. This will be the main cause of the reduction of CO_2 emission.

Table 2   Shift in annual energy demand in 2015

Option	Conventional fossil fuel	All-electric society electr.	fossil fuel	(fossil fuel in CHP also resulting in electricity)
All-electric house (540,000)	23 PJ	2 TWh		
District heating (1,500,000)	65 PJ		213 PJ	(31 TWh)
Traffic (20 percent)	41 PJ	5 TWh		
Industrial CHP (1000 MWe)	22 PJ		54 PJ	(6 TWh)
Total	151 PJ	7 TWh	267 PJ	(37 TWh)

3. ENERGY SUPPLY BY POWER GENERATING OPTIONS

The power generating options relevant in this survey have been indicated in the introduction. These options are:
- the high-efficiency combined-cycle power unit
- the standard coal-fired steam boiler/steam turbine power unit
- the integrated coal gasification combined-cycle power unit
- the combined heat and power unit based on the combined-cycle unit
- renewable energy resources.

The following sections will discuss these different options in more detail. The most relevant information on techniques, efficiencies and CO_2 figures are put together.

3.1 The high-efficiency combined-cycle power unit

The principle of the combined-cycle power unit is the integration of a gas turbine and a steam turbine. The development of this combined cycle has been very promising in recent years. This development was possible because of solid improvements of the gas turbine and by the use of more efficient and thus more complicated steam cycles. The electric efficiency of the combined-cycle units thus improved from 47 percent to 52 percent in less than 10 years. A further improvement in the near future will be made with the introduction of an organic Rankine cycle (ORC). Used as a bottom cycle the ORC means a better utilisation of the low-temperature heat.

Net electric efficiency of the combined cycle with an ORC, which is certain for year 2015, will be at least an average of 53 percent a year.

Based on the natural gas from the Slochterenfield in The Netherlands, the CO_2 emission factor is 56.4 kg/GJ for direct and indirect emissions (8). In practice the combined cycle power unit can be fired with a variety of gaseous and liquid fuels.

3.2 The standard coal-fired steam boiler/steam turbine power unit

Since it has been in use for many years, the standard coalfired steam boiler/steam turbine power unit is well-known. Further developments of this type of unit are not expected to be revolutionary. Improvements especially in environmental aspects related to SO_2, dust and NO_x will continue.

The CO_2 emission caused by the desulphurisation process is negligible, particularly for low-sulphur coals. Only very minor improvements of the net electric efficiencies are as possible. Even for the year 2015 the average efficiency over a year will be about 41 percent. For the coal currently used in Dutch power stations the CO_2 emission factor is 100 kg/GJ (8).

3.3 The integrated coal gasification combined-cycle power unit

The development of the integrated coal gasification combinedcycle power unit is in progress. In The Netherlands it has been decided to construct and operate a full-scale demo plant.

The main advantages of this type of unit are related to the environment. In comparison with the standard coal-fired unit the integrated coal gasification combined-cycle unit will have a lower emissions of SO_2, NO_x and dust and the slag will come in the form of nonlixiviatable glass. Further improvements are expected in the coal gasification processes and in the gas turbines. The net electric efficiency is expected to be 43 percent.

Since this type of power unit naturally has the same CO_2 emission factor of 100 kg/GJ for the same coal as the standard coal-fired unit, the improved efficiency will result in a somewhat lower CO_2 emission. As this new technology is just entering the demonstration phase, it is hard to give firm economic indications. But it is believed that the integrated coal gasification combined cycle unit will be at least comparable in its economics with the standard coal-fired unit. The main perspectives are for improved efficiency and lower investments.

3.4 The combined heat and power unit based on the combined-cycle unit

The principle of the combined-cycle unit has already been given in Paragraph 3.1. The combination of a gas turbine and a steam turbine is also extremely appropriate for use as a combined heat and power unit. Using the wide range of available standard gas and steam turbines, it is possible to create the best possible unit for each specific situation. Whether or not it is reasonable to use a bottom cycle depends solely on the economics.

It is hard to predict the net efficiency for any possible situation. But a firm indication is a net efficiency of 50 percent on the electric side and of 30 percent on the thermal side, as the combined heat and power unit is used in a system with space heating. In an industrial application one may think of an average efficiency in the range of 40 percent electric and 45 percent thermal. So the overall efficiency of a combined heat and power unit is higher than the efficiency of the conventional power generation. The result is saving of energy and thus less CO_2. These combined heat and power units are assumed to be fired with natural gas, so that the CO_2 emission factor is 56.4 kg/GJ.

3.5 Renewable energy resources

The renewable energy resources can be considered as environmentally benign by comparison with conventional energy resources such as fossil fuels. This is so because the renewable energy resources (i.e., wind, solar, geo or hydro) are of a physical
rather than a chemical nature. So the chemical impact on the environment, which is the cause of most problems, is nearly absent; in this respect nuclear energy is equivalent. Only some minor, indirect CO_2 emission exists. Even in the situation where biomass or waste incineration is used as an energy source, it is debatable whether this will result in an extra impact on the environment. All these renewable energy resources together are implemented as the ultimate option for generating energy with no CO_2 emission. If this simplification is accepted the renewable energy resources present the extreme situation that can be reached in a real zero CO_2 emission scenario. As the situation on renewable energy resources is quite different in each country, it is hard to be specific concerning the possible contributions of these resources to the total energy supply. But, again, this is no cause for great concern, as the

renewables are only used here to illustrate zero CO_2 emission.

4. RESULTS

The question was whether an all-electric society will lead to less CO_2 emission or not. It is clear that a fully developed all-electric society is unimaginable for the year 2015. Nevertheless, our society is becoming more electrical, the share of electricity in the overall energy package is growing steadily. These developments towards more electricity have been sketched in the previous chapters. In Chapter 2 possible changes in the energy markets are outlined. More electricity and more combined heat and power are introduced in the energy market sectors of residential areas, traffic and industry. Chapter 3 describes several options for generating more electricity and more combined heat and power.

On the one hand the developments outlined are representative of ultimate forms of introducing more electricity. On the other hand the opportunities described contain enough substance to be considered in this paper.

The results for CO_2 emission in various combinations of these markets and of the generating options are presented in Table 3. The matrix in this table presents the difference in CO_2 emissions compared with the conventional situation as used at present, without any further introduc-

Table 3   Decrease in CO_2 emissions in kton per annum.

	high eff. CC-unit	standard coal-fired	coal gasif. CC-unit	CHP unit	renewables
All-electric house (540,000)	400	-800*	-700		1,300
District heating (1,500,000)	--	--	--	3,900	--
Traffic (20 percent)	1,250	-1,450	-1,250	--	3,350
Industrial CHP (1,000 Mwe)	-	--	--	500	--
Total	1,650	-2,250	-1,950	4,400	4,650

*) - is thus an increase

tion of electricity based on combined heat and power. The figures, in kton per annum, represent the clean combination of a certain market with a particular generating option. By way of interpolation it is quite easy to calculate the CO_2 emissions for intermediate combinations.

5. REFLECTIONS

In evaluating the all-electric society at first the quantifications of the CO_2 emissions as calculated for the year 2015 are discussed first. Secondly, some qualitative remarks are made for the period following, say until 2050.

5.1 Quantitative developments up to the year 2015

- To facilitate the analysis of the figures on the decrease of CO_2 emissions, it is good to know that the overall CO_2 emission in The Netherlands was about 150,000 kton in the year 1987. So the figures of Table 3 are representing 1 to 3 percent of the total CO_2 emissions. The total reduction of a combination of different options is going up to 6 percent.
- The public power generation in 1987 and 1988 was 56.5 TWh. The related CO_2 emissions were about 36 kton, thus 24 percent of the total CO_2 emissions in The Netherlands. Generating the same 56,5 TWh by means of high-efficiency combined-cycle power units, fired with natural gas, should result in 21,000 kton of CO_2 emissions. The reduction of 15,000 kton is equivalent to 10 percent of the total CO_2 emissions in The Netherlands now (9, 10)
- The various market sectors are only mutually comparable for the specific situation in The Netherlands. In other countries these market sectors may have a completely different energy package.
- The introduction of more renewable energy resources will result in substantial reductions of CO_2 emission. However, for the introduction of more renewable energy it is not necessary to create new markets for electricity. For the time being, therefore, the renewables will not bring the all electric society.
- Also, the introduction of more combined heat and power will lead to less CO_2 emission, which is mainly because of the large saving of energy. Especially combined heat and power in district heating combines well with all-electric houses, as the energy demand for space heating occurs simultaneous.
- With respect to CO_2 emissions, it is not reasonable to change over to an all-electric society, as the intention is to use more coal-fired units, for instance during the night.

- In going all-electric, only the gas-fired high-efficiency combined-cycle power unit is helpful.
- For the national balance of payment the introduction of more combined heat and power and of more renewable energy resources is always positive.
- Since natural gas is used already, it is always wise to look for more combined heat and power and for more high-efficiency combined-cycle power units.

5.2 Qualitative developments up to the year 2050

- A further penetration of electricity until the year 2050, in the same way as described above, can result in another 5 till 10 percent reduction of the CO_2 emission.
- Generally speaking, it is to be expected that the use of fossil fuels will come under greater pressure, especially petroleum products will become rather scarce.
- Natural gas will be available in reasonable amounts. But for policy reasons this gas must only be used in an energetically (or rather exergetically) correct way. This means for instance that no gas can be used for space heating direct, but only via combined heat and power.
- Another, more sophisticated, solution can be found in the gasification of coal. One advantage of coal gasification is that the disengaged CO_2 is far easier to handle, as the CO_2 is here far more concentrated than in a standard coal-fired power unit. Another benefit of the coal gasification process is that it also produces H_2 in a usable form. This H_2 can be used for instance in fuel cells, for traffic and for mixing with natural gas.
- Among the fossil fuels coal will play a more important role. However, the introduction of more coal will result in more CO_2 emission. A solution to this problem can be found in more combined heat and power.
- As our energy is currently mainly derived from non-renewable and exhaustible resources, our task for the future is to save energy, which will be done by means of combined heat and power and renewable resources.
- On the demand side the energy market will continue to change, the trend will remain unchanged, which means less energy and relatively more electricity.

6. CONCLUSIONS

This paper has only focused on the effect of a more electric society effect on CO_2 emission. In other words, no other environmental aspects of energy supply have been discussed. From the point of view of CO_2 emission only the following conclusions can be drawn:

- our future society will use less energy and relatively more electricity, which will results in electricity also penetrating new sectors of the energy market
- a better use of fossil fuels is necessary, which clearly calls claim for more combined heat and power
- more and continuous attention must be given to renewable energy resources
- as coal is an important resource for the future, the development of the coal gasification option is highly important
- an all-electric society is not a guarantee for less CO_2 emission, but the electric part of our energy supply can and indeed must must play a role in the decrease of CO_2 emission.

7. REFERENCES

1. Energy Statistics Yearbook 1985, United Nations
2. CBS - Energiestromen in Nederland 1987, Central Office of Statistics of the Netherlands
3. De gasloze woning (KEMA-study 1984)
4. Distributie 2010 (KEMA-study 1986)
5. All-electric wonen 2010 (TUE/KEMA-study 1989)
6. VESTIN Jaarverslag 1987
7. Uitgangspunten en achtergronden voor NOVEM-studie "Kansen voor alternatieve brandstoffen in het wegverkeer", ESC-WR-88-25 (1988)
8. CO_2 Emissies bij verschillende alternatieven van elektriciteitsopwekking en -besparing in Nederland, ESC-WR-89-11 (1989)
9. Elektriciteit in Nederland 1987
10. Elektriciteit in Nederland 1988

HYDROGEN AS AN ENERGY CARRIER TO REDUCE CO_2 EMISSIONS

H. MUIS and A.F.L. SLOB
Consultants on Energy and the Environment (CEA)
P.O. Box 21421
3001 AK Rotterdam
The Netherlands

ABSTRACT.Our present energy system, based mainly on the extensive use of fossil fuels, is convenient and cheap, but also has many adverse environmental effects. Furthermore, resources of these fuels are limited. A new energy system, using both electricity and hydrogen as secondary energy carriers, may provide an environmentally sound link between primary energy sources and final demand. For its primary sources the system makes use of renewable energy and, on a temporary basis, fossil fuels. Although the costs of such a hydrogen system will be substncially higher than our present energy system, it probably will also be the best way to ensure a sustainable future. The system as described below, avoids pollution by carbondioxide and other emissions and is not vulnerable to depletion.

1. INTRODUCTION

Today, almost all world energy demand is met by fossil fuels. Especially liquid fossil fuels have some very useful properties: they possess energy in a very concentrated form, they can easily be transported using pipelines, tankers and trucks and they can be stored for a long time without losing their properties. At the same time, however, they are an exhaustible source of energy, bringing billions of tons of CO_2, CO, SO_2, NO_x, soot and ash into the atmosphere every year. Nor are fossil fuels evenly distributed around the world, which makes them a potential cause for international conflicts.

Renewable forms of energy (solar, wind, hydropower), on the other hand, do not possess the above-mentioned shortcomings of fossil fuels, but they lack the advantages of easy handling, storing and using.

To overcome the shortcomings of both fossil based and

renewable energy systems while keeping the advantages, scientists and engineers have in the last decades proposed the hydrogen energy system. In such a system hydrogen is not a primary energy source but is used as a carrier and storage medium. Several primary energy sources, both fossil and non-fossil, can be used to produce the hydrogen. Hydrogen forms the link between these sources and the user, where it can be converted into heat, traction or electricity with very high efficiency and almost no emissions.

The idea of an energy system based on hydrogen is not completely new. More than a hundred years ago, Jules Verne envisioned water, decomposed by electricity into hydrogen and oxygen, as the "coal of the future".

Until the beginning of the 19th century "inflammable air" as it was called, only found limited application in balloons since it was lighter than air.

In the 19th century the first energy uses of hydrogen were applied. In 1807 the Pall Mall in London was illuminated by hydrogen lamps. From 1820 on more cities applied this method of street illumination. Some decades later hydrogen was also used for heating purposes (space heating, cooking). All hydrogen until that time was produced in a chemical reaction between iron and steam to form iron oxide and H_2. Around 1850 a new process was discovered: a reaction between coal and steam producing a mixture of hydrogen and carbonmonoxide. In most cities large installations were erected to produce this so-called town gas which was widely used for heating, cooking and illumination until only some decades ago.

In the period 1910 - 1940 another method of producing hydrogen came into use, after the invention of the electric generator: electrolysis of water. In the 1960s there still existed a worldwide capacity of about 2.000 MW for producing hydrogen electrolytically, mainly using hydropower (Norway, India, Egypt).

Meanwhile, however, due to the increasing demand for hydrogen in the chemical industries, other production methods were invented using fossil fuels (coal, oil, gas) which were cheaper where no cheap electricity was at hand.

But since natural gas, oil and electricity could meet the demand for heating, cooking and illumination more cheaply or more conveniently, the use of hydrogen as an energy carrier diminished and nowadays, almost all of the annually produced 30 million tons of hydrogen is used in the chemical and refinery industries. About 80% of this amount is produced from methane, being the cheapest process and only 1% is produced via electrolysis.

It was not until around 1970 that hydrogen as an energy carrier regained attention, because it was felt that fossil fuels would not last forever. The idea was to produce hydrogen by means of renewable energy or nuclear energy (fission). The nuclear reactors could be located far away from population centres for reasons of safety while the energy would be transported in the form of hydrogen to the cities. For renewable energy sources the same argument applied while at the same time the energy would be stored to overcome the natural fluctuations. This is the time when the first theories about the hydrogen society emerged: an infrastructure based upon hydrogen and electricity, one form of energy converted into the other by means of electrolysis or fuel cells with very high efficiencies (80-90%), the primary sources being renewables or fission. Hydrogen and electricity could provide all the necessary end uses. This concept promised a never ending flow of energy. Over the last decade, however, it has become increasingly doubtful whether the nuclear option is realistic in the long run and more emphasis is put on the environmental benign aspects of the proposed hydrogen energy system.

Hydrogen production with renewable sources produces no emissions. Storage and transportation have no adverse effects while the end use in heating, electricity production or traction has only minimal emissions of NOx, depending on the temperature. In table 1 some properties of gasoline, natural gas and hydrogen are given. It appears that hydrogen can perform much the same functions as gasoline or natural gas.

Table 1. properties of gasoline, natural gas and hydrogen (Veziroglu, 1987)

Property	Gasoline	Natural gas	Hydrogen
Density (kg m^{-3})	730	0.78	0.084 0.71×10^{-1} (liquid)
Boiling point (°C)	38/204	- 156	- 253 (20 K)
Lower heating value: gravimetric (kJ kg^{-1}) volumetric (kJ m^{-3})	4.45×10^4 32.0×10^6	4.8×10^4 37.3×10^3	12.50×10^4 10.4×10^3 (gas) 8.52×10^6 (liquid)
Flammable limits (% in air)	1.4-7.6	5-16	4-75
Flame speed (ms^{-1})	0.40	0.41	3.45
Flame temperature in air (°C)	2197	1875	2045
Ignition temperature (°C)	257	540	585
Flame luminosity	high	mediu	low

As with these two fuels, hydrogen, too, needs special attention regarding safety. Like all other fuels, hydrogen can burn catastrophically as the cases of the Hindenburg and the Challenger have proved. It will not explode easily, however, since a certain volume of air must contain more than 75 vol.% of hydrogen before the explosion limit is exceeded. This limit, therefore, lies much higher than with other fuels. In actual practice it will be difficult to reach such a high concentration because hydrogen, as a result of its small dimensions, diffuses very rapidly. When, in addition, the high ignition temperature and the low luminosity of a hydrogen flame (little radiation of heat) are taken into account, hydrogen may be considered as a relatively safe fuel.

2. PRESENT STATUS OF HYDROGEN AS AN ENERGY CARRIER

At present the only energy use of hydrogen of any significance is in the burning of cokes oven gas which contains about 50% of hydrogen. But this is not a case where hydrogen can be said to be specifically produced for its favourable environmental aspects.

Although today there is as yet little evidence of the above mentioned hydrogen society, all the technologies needed for this concept are available and have been demonstrated. Electrolysis is a proven technology, storage and transportation is state-of-the-art (in the Federal Republic of Germany a pipeline for hydrogen exists with a length of several hundred kilometres), cars, airplanes and spacecraft have been propelled with hydrogen, heating appliances have been made and fuel cells which convert hydrogen into electricity with an efficiency of 60% are in use. All elements are ready, waiting for the moment the hydrogen energy system will be competitive with the existing energy system or simply waiting until the adverse environmental effects of the fossil fuel energy system will become prohibitive.

Before going into the details of an energy system with little or no environmental damage, the elements of the hydrogen system will be discussed.

2.1 Production of hydrogen

Six different methods for hydrogen production are available at present, only two of which are currently in use on a commercial scale.
- production from fossil fuels
- electrolysis of water
- thermochemical processes
- photochemical dissociation of water
- direct dissociation at very high temperatures
- biological methods

Of the present hydrogen production 80% is produced by steam reforming of methane (CH_4 + $2H_2O$ $\rightleftharpoons$ CO_2 + $4H_2$). Other methods are partial oxidation of oil and gasification of coal. All of these methods produce emissions of aerosols and soot, SO_2, NO_x, CO and CO_2. Steam reforming of methane is the cheapest and cleanest process. Hydrogen produced in this way costs about Dfl 15,=/GJ (1986) while coal gasification yields hydrogen at Dfl 20,=/GJ. It is believed that some years after 2000 hydrogen from coal will become the cheapest method of production because of the large reserves of coal. A disadvantage of coal gasification, compared to steam reforming of methane, is that, because of the low hydrogen content of coal, much more carbondioxide is formed per amount of hydrogen produced.

The technology of electrolysis of water has long been known. In 1900 the first large scale installations, located in Norway and Iceland, operated with hydropower. The process is called conventional electrolysis: two electrodes are placed in a 25% KOH solution through which an electrical current is fed. At one plate oxygen is formed, the other produces hydrogen. Both gases are kept separate by means of a membrane. A number of such cells are stacked to form large arrays. About 4 kWh is needed to produce one m^3 of hydrogen gas. Over the last decade electrolysers were built which operate at higher currents (0.5 A/cm^2), higher temperatures (120 °C) and higher pressures (5 bar): this is known as advanced electrolysis. Efficiency is about 85%. When cheap electricity is available (from hydropower or off-peak nuclear energy) the costs of the produced hydrogen are comparable with costs involved in steam reforming.

The thermochemical process for hydrogen production makes use of several reactions in which the energy, necessary to split the water molecule into its constituents, is transferred. Although several routes have been proposed and some have even been demonstrated, the cheap high temperature source needed for the reactions (a high temperature gascooled nuclear reactor or a solar power tower) is not yet available and maybe never will be. Besides, the necessary hazardous chemicals involved in the process pose a significant environmental risk.

The process of photochemical dissociation of water (or photo-electro-chemical) is in fact a combination of photovoltaic electricity generation (PV) and electrolysis at the same time. It is more difficult than normal PV and subsequent electrolysis and does not have a higher efficiency but hopefully it will prove cheaper in the end.

Direct dissociation of water at very high temperatures (> 2.500 K) is at present only a theoretical option.

Biological methods, employing bacteria, which may eventually be genetically engineered, or other micro organisms, is also a route to hydrogen production which still needs a great amount of research. The perspective is to convert waste, dung and other organic material into clean hydrogen.

Table 2 presents some figures of production costs of hydrogen, synthetic natural gas (SNG) and synthetic gasoline (SynGas) by 2000, based upon recent literature (Veziroglu, 1987).

Table 2: production costs of some synthetic fuels

Synthetic fuel	Estimated average gaseous fuel cost (1986 US$ GJ^{-1})	Estimated average liquid fuel cost (1986 US$ GJ^{-1})
Coal H$_2$	7.19	8.99
Hydropower H$_2$	9.65	12.06
Other	18.13	22.66
(solar, wind, nuclear etc.) H$_2$		
Synthetic fossil fuels	7.42 (SNG)	14.47 (SynGas)
(gas/gasoline)		

2.2 Storage and transport

Hydrogen can be stored or transported in one of three forms: gaseous, liquid or attached to another material.

Storage and transport in gaseous form is much the same as with natural gas. It can be stored in pressure vessels, and in underground cavities, depleted gasfields or aquifers (geological anticlinal pourous structures). The type of storage that is planned by Gasunie for the storage of natural gas (solution mined caveties in a salt dome) could also be used for hydrogen. Transport is the same as for any other gas, in fact the former town gas contained 50% hydrogen. Attention, however, has to be paid to the risk of hydrogen brittling (steel can become brittle as dissociated hydrogen enters the crystal structure and weakens the material).

Storage and transport in liquid form is also much the same as with liquid natural gas (LNG), although the boiling temperature of hydrogen is somewhat lower. In the NASA Shuttle programme liquid hydrogen (and liquid oxygen) is used for propulsion. A third form of storage and/or transport is when the hydrogen is chemically bound to another molecule or alloy. The method has been demonstrated with quite a number of metal alloys. The reaction takes the form: $2Me + xH_2 \rightleftarrows 2MeH_x + \Delta Q$. Heat applied to the hydride, formed in this way, will release the gaseous hydrogen again. Per unit of volume even more hydrogen can be 'stored' thus than in liquid form, although expressed in kilogram of storage the amount of hydrogen is only 1 - 5%.

2.3 End use of hydrogen

Hydrogen, as an energy carrier, can be used for heating (space heating, cooking, hot water), electricity generation and transportation.

Space heating with a very high efficiency (close to 100%) is possible by using wall panels where hydrogen is catalytically combined with oxygen at low temperatures. No emissions, no chimney. Cooking can also be done using catalytic combustion, the most important advantages being the lack of a flame and the low temperature which permits a very low production of NO_x. Hydrogen can of course also be burned in the conventional way for space heating and domestic hot water.

Once hydrogen is produced, it is a very efficient fuel for fuel cells. In actual practice fuel cells can have efficiencies of 60% while in theory 80 or 90% seems a possibility. They contain little moving parts, produce very little noise and no emissions. The heat which is produced at the same time, can be used for heating purposes. In Tokyo a 4.5 MW phosphoric acid system was built in 1985. Many types of fuel cells have been developed using, except pure hydrogen, all kinds of hydrogen containing fuels (Kordesh, 1988). Prototype fuel cells today costs about $2000-$3000 kW^{-1}. It is expected mass produced fuel cells will cost only about $350 kW^{-1}.

Another way of using hydrogen is in traction. Hydrogen can be burned in Otto-engines and even in Diesel-engines, in a gaseous form, liquid or mixed with gasoline. The practical use has been extensively demonstrated in the FRG during the last decade. The feasibility of an infrastructure with liquid hydrogen has also been demonstrated (Feucht et al., 1988; Peschka, 1987). Hydrogen is also considered for the propulsion of airplanes. Since it has the highest energy content per unit of weight, the weight during take-off will be minimized. The airport in Zürich has studied the feasibility of using liquid hydrogen but concluded it would still be 2 - 4 times more expensive than conventional jet-A fuel (Alder, 1987). The use of hydrogen in airplanes instead of conventional fuel would greatly reduce pollution in the higher atmospheric regions since only water would be emitted. Although hydrogen can be used directly in internal combustion engines, conversion into electricity via fuel cells and subsequent conversion into mechanical energy with an electric motor will yield much better overall efficiency. The only problem is that the volume needed in the latter case is much larger so the first applications can be expected in buses and vans. Making use of fuel cells, the efficiency is at least twice as good as with internal combustion engines. If hydrogen is burned directly in combustion engines, performance is still about 25% better than when gasoline is used.

3. THE HYDROGEN CYCLE

As has been shown in the previous section, all the elements for a hydrogen energy system are available. Facing the increasing threats to the environment, caused by our conventional infrastructure with its predominant use of fossil fuels, it is interesting to design an energy system, based on hydrogen and electricity, which does not have these adverse environmental effects. In figure 1 the concept is given of such an energy infrastructure. Transmission of energy over very long distances will be cheapest in the form of hydrogen (for distances longer than approximately 600 km it even pays to convert electricity into hydrogen and after transmission reconvert into electricity again). Forintermediate distances electric transmission lines can be used and only for very small distances (e.g., < 5 km) transmission of hot water is feasible.

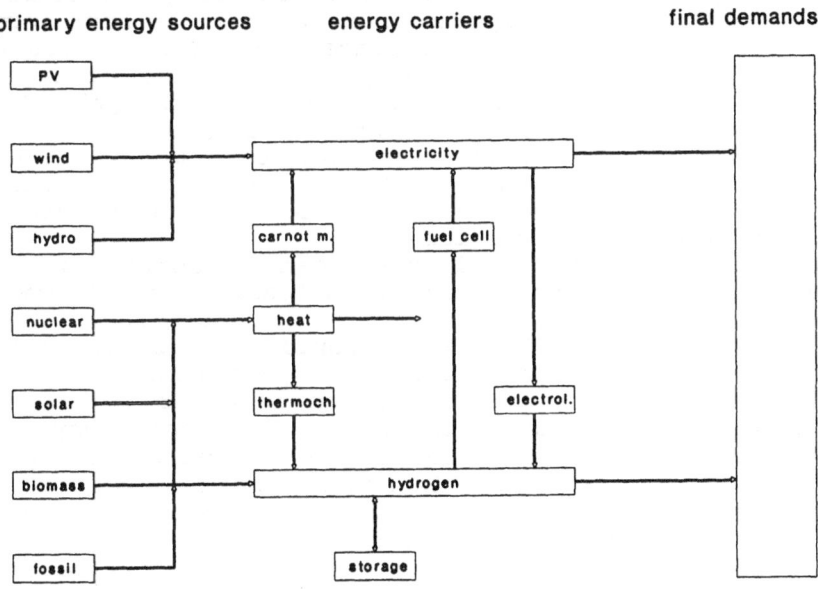

Figure 1: concept of the energy system in a hydrogen society

With regard to emissions of CO_2 two options have to be considered depending on which primary energy sources are chosen to feed the system: renewable energy sources or hydrocarbons.

When renewable sources are used (sun, wind, hydropower), no emission of CO_2 occurs in any part of the system: production, storage and transport, final use. There will be no emissions whatsoever, except for some NO_x perhaps in the final use, when no or insufficient use is made of suitable catalysts to keep the temperature low. And, of course, the end product of hydrogen combining with oxygen will be released: water of high purity. Other possible environmental problems associated with a hydrogen system fed on renewables are concerned with safety aspects (handling of the hydrogen, wind turbines) but these are well known and can be controlled, and there may be problems associated with hydropower (inundation, loss of biotope, modification of flow). Then there is the possible problem of noise which can be generated by wind energy conversion systems (WECS). All of these possible risks or problems, however, play a role only on a local or regional scale and do not play a part on a continental or global scale.

The second option to be considered is when fossil fuels (methane, oil or coal), biomass or nuclear energy are used as the primary energy sources. In the long run nuclear energy is not a feasible option (nor is it on a short term) for the production of hydrogen since the resource base of uranium is limited and fusion energy is still out of sight. So fossil fuels and biomass will have to do the job. Coal will be abundant for some centuries to come while the reserves of gas and oil are expected to come to an end somewhere in the next century. Biomass, on the other hand, can be a truly renewable primary source, if proper care is taken of reforestry and the like.

Both biomass and fossil fuels can be converted into hydrogen and CO_2 by partial oxidation (gasification), where carbonmonoxide is formed, followed by the shift reaction $CO + H_2O \rightleftharpoons CO_2 + H_2$. The carbondioxide must be separated from the hydrogen and should then be injected into depleted gasfields, aquifers or the deep ocean in order to prevent their release into the atmosphere. The technology for gasification, separation of gases and injection into gasfields or aquifers is available and has been demonstrated. The main advantage of converting hydrocarbons into hydrogen in large central plants, instead of producing synthetic gases or fuel, is that all unwanted emissions can be captured and disposed of in an environmentally sound way. Except for the production of gases and fly ash as mentioned before, there will also be, where coal is used, quite large amounts of ash which may have to be shipped back to the mines. Other environmental problems associated with the use of biomass or fossil fuels are the risk of polluting rivers, seas and oceans (oil), destruction of landscape (coal) and problems related to the transportation of the fuel to the plant (noise, heavy traffic, ship movements).

The first option, where only renewable energy is used, clearly has the edge over the second, from the environmental point of view. It may be expected, however, that at the time the hydrogen society will be a fact, both options will have been realised simultaneously, presumably with the initial use of fossil fuels and then a gradual shift to more and more renewable energy sources.

Another point which has to be made is that the scale of a hydrogen energy system may be national but will probably grow to be continental since economies tend to get more interrelated (EG, internal market, détente). The supply of renewable energy sources is more stable on a larger scale (diversification: hydro in Norway, solar in the Mediteranian or the Sahara) and the transport of hydrogen over long distances poses no problems.

It should be noted that a hydrogen society is highly capital intensive, as is our present energy system. In the Netherlands, for instance, about 10.000 km. of pipelines for transport (costing Dfl. 6 billion) and 80.000 km. of pipelines for distribution (also costing some Dfl. 6 billion) is presently in use for the distribution of 35 billion m^3 of natural gas yearly to 5 million end users. For developing countries it will therefore hardly be feasible to adopt such an infrastructure in the short term, or even in the intermediate term. So, on a global scale the release of CO_2 because of decentralized burning of fossil fuel will continue. For the developed countries, however, it seems even possible to actually decrease CO_2 concentration in the atmosphere when biomass is used and the resulting CO_2 is injected into the underground. On a temporary basis the same can be accomplished by growing more forests than are cut down.

4. IMPLEMENTATION OF THE HYDROGEN CYCLE

There are several reasons for the continued use of fossil fuels in large amounts: modern society depends on them, they are convenient and they are cheap (in this order). The use of energy in several forms is intimately connected to our way of living. There is, however, still a large quantity of energy which can be conserved. The rate at which energy conservation takes place is not very impressive though, due to the fact that fossil fuels are relatively cheap. And fossil fuels, especially in liquid or gaseous form, are very easy to handle. So a shift to hydrogen is not felt as an urgent need.

Still, there are some excellent reasons for making

this shift soon. We have already mentioned:
- the depletion of non-renewable energy sources, except
  coal
- the environmental problems on all possible scales (local,
  regional, fluvial, continental, global)
It is also important to note that every major new energy
carrier needs 40 - 50 years before it contributes its first
10% share of world energy supply.

Obviously, the main obstacle for a quick introduction of
hydrogen as an energy carrier is its price, in relation to
that of fossil fuels. Table 3 gives an overview of the
costs involved for some methods of production and the
amount of primary energy needed. This price consists of two

Table 3: fixed costs, primary energy needed and total costs
of hydrogen (1986)

process	fixed costs Dfl./GJ	primary energy	cost of primary energy	total cost Dfl./GJ
steam reforming of methane	3,50	1,30 GJ/GJ	Dfl.9,00 /GJ	15
coal gasification	12,50	1,80 GJ/GJ	Dfl.4,00 /GJ	20
advanced electro-lysis	3,50	292 kWh/GJ	Dfl.0,10 /kWh	33

To obtain the hydrogen in liquid form about 25% must be added to the
stated prices.
Adequate disposal of CO_2 may add about Dfl. 1,= to the total cost per
GJ.

parts: a fixed amount, accounting for capital investment
and maintenance, and an amount for the primary energy which
is converted into hydrogen. As these figures clearly show,
hydrogen derived from fossil fuels, or from electricity
generated by burning fossil fuels, is considerably more
expensive than the original fossil fuels. The conversion of
natural gas into a form (hydrogen) where no pollution in
the final use occurs, will add about 50% to its original
costs (the disposal of CO_2 is not yet accounted for in
these figures). It is also clear that if electricity is
used to produce hydrogen from water, it has to be really
inexpensive in order to be competitive with hydrogen from
natural gas or coal. Nowadays only off-peak hydropower or
nuclear power can be purchased at such a low price.
    So, if hydrogen from renewables is to compete with
hydrogen from fossil fuels, or, more specifically from

coal, electricity generated by these renewables should not cost more than approximately Dfl 0,06/kWh, that is, when the disposal of CO_2 and other gases in aquifers does not add substantially to the costs of hydrogen from coal. In fact, as a recent study indicates (Krekel, 1987), within a number of decades the price of coal may increase to Dfl. 5 - 8,00/GJ whereas electricity from renewables may reach prices as stated before (around Dfl. 0.06/kWh).

It will take some decades before hydrogen produced from renewables will be competitive with hydrogen produced from coal, with a price stabilizing somewhere between Dfl. 20,00 and Dfl. 30,00 per GJ. This is two or three times the price we pay today for our liquid or gaseous fossil fuels. Comparing the costs of fossil fuels and hydrogen it must be borne in mind that in most cases the conversion efficiency of hydrogen is higher.

Sometimes, the argument is reversed. It is argued that fuel prices should be two or three times as high as they are now, because at present they do not account for the costs of damage to human health, wildlife and ecosystems. Tentative calculations of these so-called social costs have come up with a figure of around Dfl. 20,00/GJ (Veziroglu, 1987; Hohmeyer, 1988), which may be the very price one has to pay for an energy system which is ecologically benign.

History shows how a great many older energy systems were replaced by new ones, not because the new technology was cheaper but just because it was better. For instance, coal was replaced by natural gas for heating purposes although gas is more than twice as expensive. Electric lighting has replaced gas lighting (even electric cooking has replaced gas cooking in a substantial number of families) not because electricity is cheaper (in fact it is three times as expensive as gas), but because it is more convenient or cleaner. The same argument, now on a global scale, applies to hydrogen.

Hydrogen poses no major problems, apart from its production costs. The necessary infrastructure for a hydrogen energy system already partly exists: there is a system of electric transmission lines and there is an underground piping system throughout the Netherlands (now in use for natural gas) which may be used for hydrogen. So, the higher costs for hydrogen at present is the price that has to be paid for the avoided emissions, mainly carbondioxide. In the case of domestic heating, for instance, using hydrogen, produced through steam reforming of methane and injecting the CO_2 into the underground, instead of natural gas itself, will cost about Dfl. 115,- per ton of avoided CO_2 (see table 3, emission of CO_2 for natural gas is 56 kg/GJ, injection costs Dfl. 7.50/ton). In traction the costs of avoiding CO_2 release into the atmosphere will be approximately Dfl. 50-70,- per ton CO_2 (assuming emission of CO_2 for gasoline is 73 ton/GJ, gasoline costs Dfl. 9,-/GJ, gaseous hydrogen is used, no extra costs for equipment). These calculations must be considered as tentative.

In table 4 gross and nett final energy demand in The Netherlands is given for 1990 (Krekel, 1987). The difference between gross and nett demand is caused by conversion losses between primary sources and the final energy carriers. As can be seen, about 140 million tons of carbon

Table 4. yearly emission of carbon dioxide in The Netherlands (1990)

type of end use	gross demand million GJ	conv. eff.	nett demand million GJ	CO_2 release million tons
electricity	585	0.4	234	40
traction	525	0.85	446	30
heat	1190	1	1190	70
feedstock	276	–	276	–
total	2576		2125	140

dioxide is released yearly. In a hydrogen society all final demands will be met by hydrogen or electricity. For some applications hydrogen will be the best energy carrier, other energy demands are better met by electricity. For traction hydrogen seems the best way to prevent CO_2 emissions. The demand for heat can be met both by hydrogen and electricity. It takes a thorough systems analysis to determine the optimal size and shape of such a sophisticated clean energy system, including further possibilities for energy conservation. In case no other measures are taken and all energy demand in both traction and heating is supplied by hydrogen, yearly 100 million tons of carbon dioxide emissions may be avoided.

In figure 2 an indication of yearly extra costs is given.

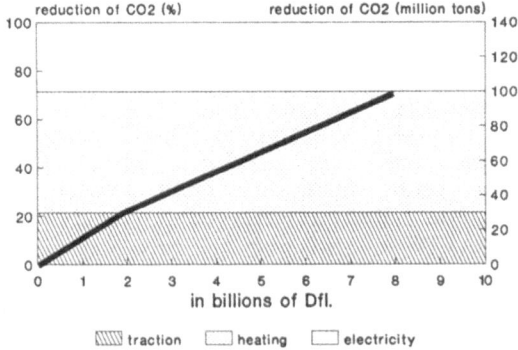

Figure 2: an indication of yearly costs of avoiding CO$_2$ emission by using hydrogen

The amount of avoided CO$_2$ and the costs are directly proportional to the penetration of hydrogen in these areas.

In reality there may be limitations in the production of the hydrogen. In The Netherlands, for instance, it may be impossible to base such a system completely on renewable energy sources. In fact, renewables are thought to supply only about 10 - 20% of the demand by 2050, another 20 - 30% will be provided for by energy conservation while 50 -70% will still be supplied by fossil fuels (mostly coal) (Krekel,1987).

This applies to a situation where energy is still considered a matter of national interest only. When energy systems are designed on a continental scale, however, more possibilities for renewable energy sources arise. On an international scale the arid, sunny regions which are now most often extremely poor, could export their sunshine in hydrogen form, becoming at the same time a participant in the global trade system.

The rate at which hydrogen will penetrate as an energy carrier into our society depends almost entirely on how the environmental problems associated with our present energy system are really considered urgent. If hydrogen is to become an important energy carrier, it will be so, not because it is cheaper, but simply because it is better. In the meantime we have to look for niches in the energy market where environmental problems are considered severe and society is willing to pay more for clean air.

An example is city traffic. On hot days the exhaust gases of cars (combined with other industrial and agricultural emissions) cause problems in the larger cities (smog). The problem can be tackled by decreasing private transport and increasing public transport, like trains and tramways using electricity or buses using hydrogen.

Some expect the first breakthrough of hydrogen as an energy carrier in air transportation (Marchetti, 1987).

Another application which is almost feasible is in using fuel cells for decentralized cogeneration of heat and electricity. The high efficiency of the fuel cells compensates for the higher costs of the necessary hydrogen.

Burning hydrogen with pure oxygen yields a very compact steam generator which can be used to produce additional steam almost instantly during peak demands. This steam can be injected into a steam turbine. Prototypes of this system have been demonstrated in the FRG.

It is important to note that the change to a hydrogen society does not have to be abrupt. Quite a number of technologies can be used with traditional fuels first, with a gradual shift toward the use of hydrogen. An example is the national pipeline system for natural gas: natural gas can be mixed with up to 20% of hydrogen without changing its properties. So, the policy to accelerate the implementation of the hydrogen system is to encourage applications which are already economically feasible or nearly so and to increase the price of fossil fuels in those areas where environmental problems are felt to be most urgent (or to subsidize hydrogen applications in those areas).

5.   CONCLUSIONS

An energy system, based on hydrogen and electricity as the main energy carriers, can provide for all forms of final energy demands. Hydrogen can be produced from renewable energy sources with no emissions of CO_2.

Hydrogen can also be produced from fossil fuels, in the short term from natural gas or oil, in the long term from coal and biomass. The resulting CO_2 can be injected into underground reservoirs.

In a hydrogen society no relevant emissions will occur, except for water.

All necessary elements of a hydrogen-based energy system are already available or have been demonstrated. For several decades to come, a hydrogen system will not be cheaper than a fossil based energy system, unless environmental costs will be fully accounted for.

The main reason for the implementation of hydrogen in the energy system is the fact that it is (almost) pollution free. Since new energy systems need several decades before they can supply a substantial share of the final energy demand, the implementation of hydrogen should start soon, to begin where environmental problems are most severe.

In order to accelerate the implementation of hydrogen, the price of fossil fuel applications could be raised for those areas where the adverse environmental effects of CO_2 and

other emissions are felt most; alternatively the application of hydrogen in those areas could be subsidized.
Promising applications for hydrogen as an energy carrier in the short term are the use in public transport and in fuel cells for the generation of both heat and electricity.

REFERENCES

Alder, H.P. (1987). 'Hydrogen in air transportation, feasibility study for Zürich airport, Switzerland', in Hydrogen Energy, 12, pp.571-585.
Feucht, K., W. Hurich, N. Komoschinski and R. Povel, (1988). 'Hydrogen drive for road vehicles -results from the fleet test run in Berlin', in Hydrogen Energy, 13, pp.243-250.
Hohmeyer, O. (1988). Social Costs of Energy Consumption, Springer Verlag, Berlin.
Kordesch, K., and J.C.T. Oliveira, (1988). 'Fuel cells: the present state of the technology and future applications, with special consideration of the alkaline hydrogen/oxygen (air) systems', in Hydrogen Energy, 13, pp.411-427.
Krekel, N.R.A., P.A.M. Berdowski and A.J. van Dieren, (1987). Duurzame energie, een toekomstverkenning, KWW, Rotterdam.
Marchetti, C. (1987). 'The future of hydrogen -an analysis at world level with a special look at air transport', in Hydrogen Energy, 12, pp.61-71.
Muis, H and A.F.L. Slob, (1987). De inzet van waterstof in de energie voorziening, stand der techniek en ontwikkelingen tot 2050, CEA, Rotterdam.
Peschka, W. (1987). 'The status of handling and storage techniques for liquid hydrogen in motor vehicles', in Hydrogen Energy, 12, pp.753-764.
Slob, A.F.L. and H. Muis, (1986). Waterstof, een verkenning van de stand der techniek, CEA, Rotterdam.
Veziroglu, T.N. (1987). 'Hydrogen technology for energy needs of human settlements', in Hydrogen Energy, 12, pp.99-129.
Winter, C.-J. (1987). 'Hydrogen energy -expected engineering break throughs', in Hydrogen Energy, 12, pp.521-546.

THE CARBON-DIOXIDE SUBSTITUTION POTENTIAL OF METHANE AND URANIUM RESERVES

Bert de Vries
IVEM - Center for Energy and Environmental Studies
State University of Groningen
P.O. Box 72
9700 AB Groningen, The Netherlands

ABSTRACT. Methane and uranium have the potential to reduce CO_2 emissions from energy use. For two sets of scenarios, the reduction potential is estimated in relation to cumulated natural gas and uranium-oxide (U_3O_8) requirements. Some methodological issues in reserve estimation are discussed; next, an overview is given of proved and unproved as well as speculative reserve estimates of methane and uranium. The geostatistical approach is discussed in some detail to highlight the nature of inference from known deposits onto unknown reserves.

1. INTRODUCTION

In recent years, with the realisation of the probable climate change due to an increasing atmospheric CO_2 concentration, it has been pointed out that substitution of coal and oil for fuels with a lower CO_2 emission may mitigate the consequences, or at least the rate of occurrence [1,2]. One obvious question, then, is to assess the long term availability of such low-CO_2 emission fuels. In the present paper, I will address this question for the two major substitution options: methane and uranium. First, I indicate how important such substitution might be as part of a CO_2 emission reduction strategy and how much methane and uranium are required. Next, I assess the past and present reserves in terms of geological existence and economic availability. Finally, some conclusions are drawn as to the feasibility of a long-term fuel substitution strategy as part of dealing with the greenhouse problem.

2. FUEL SUBSTITUTION SCENARIOS

2.1 The CO_2 substitution coefficient

The first part of the question is, how effective methane and uranium are in stabilizing or even lowering global CO_2 emissions. To indicate this issue I have made some forward calculations. Starting point is the assumption that energy end-use will remain at the present level. Consequently, I express CO_2 emissions in units per unit of end-use fuel and electricity. End-use stabilisation appears an appropriate assumption: some argue that it is impossible in view of the developing worlds' needs, others feel it is the minimum target in avoiding major ecological catastrophe. End-use stabilisation can be realized through all forms of increasing energy productivity e.g., insulation, cogeneration, solar roofs and the like. The resulting demand for fuels and electricity has to be met from the energy supply system.

Let the present end-use be supplied from fuel shares F_0; let there be some set of target fuel shares F_T. Then, I define the CO_2 substitution coefficient σ (F_0, F_T) as the percentage reduction of cumulated CO_2 emission per unit of end-use energy in the transition from fuel share vector F_0 (present) to fuel share vector F_T (T years from present). The reference is $F_T = F_0$ ($\sigma = 0$). Given T and the assumption of linear interpolation between F_0 and F_T, one can calculate σ (F_0, F_T) and the cumulated methane and uranium requirements.

It should be noted here, that the contribution of methane to climate change in its role of greenhouse gas is not considered in this paper [3]. Nor is a possible transition to fuels of even lower C/H-ratio e.g. hydrogen, considered.

2.2 CO_2 reduction through substitution: Western Europe

Table 1a indicates the effectiveness of substitution for Western Europe (EUR-12) for various target fuel share vectors F_T. At present, in 1988, about 30.76 EJ (excluding feedstock and bunkers) has been supplied to end-use applications. Of this energy flow, 5-6 EJ has been used in the form of electricity, the remaining 82% largely being used for process heat, space heating and motive power for transport. A major assumption in Table 1 is, as stated before, that EUR-12 energy end-use will remain constant over the next decades. Another important assumption concerns the quality of end-use, for which the share of electricity is the major indicator. Due to increasing applications for electricity and the generally lower marginal cost for heat conservation, this share is expected to rise. There is, however, a considerable potential for electricity conservation too. Finally, direct and indirect specific CO_2 emission coefficients (kg/GJ) have to be chosen. Based on various values in the literature, I have selected the values in Appendix A as representative and given a margin of uncertainty. One problem in establishing future indirect CO_2 emission coefficients is that one has to make an assumption about the average CO_2 emission coefficients-which in turn depends on the fuel share scenarios and will also change with the use of more marginal resource deposits and more advanced technology. I shall deal with this issue by using two natural gas and three uranium fuel categories.

TABLE 1a CO_2 substitution from methane and uranium in Western Europe

Fuel type	Fuel share F_o	Fuel share scenarios F_t				
		A	B	C	D	E
Natural gas (heat, trp) cost category I	0.218	0.287	0.287	0.548	0.274	0.548
Natural gas (heat, trp) cost category II	0	0	0	0	0.274	0
Natural gas for electricity	0.013	0.026	0.026	0.052	0.052	0.052
Uranium low cost 0.1%	0.067	0.067	0.137	0.198	0.066	0.121
Uranium medium cost 0.01%	0	0	0	0	0.066	0
Uranium high cost 0.004%	0	0	0	0	0.066	0
Light FOP, LPG (heat, trp)	0.403	0.38	0.38	0.132	0.132	0-18
Heavy FO (heat, trp)	0.09	0.067	0.067	0.03	0.03	0.051
Heavy FO for electricity	0.02	0.02	0.02	0	0	0
Coal (heat, trp)	0.045	0.022	0.022	0	0	0
coal for electricity	0.073	0.06	0.030	0	0	0
Other fuels (heat, trp)	0.002	0.002	0.002	0.002	0.002	0.002
Hydro and other electricity	0.009	0.009	0.009	0.009	0.009	0.009
Metallurgical coal	0.06	0.06	0.06	0.03	0.03	0.03
Anticipation period T (yr)	-	30	30	60	60	60
Avg annual CO_2-emission (Gton)	2.78	2.72	2.61	2.29	2.39	2.35
CO_2-substitution coefficient	-	0.022	0.059	0.177a)	0.141	0.155
CH_4-requirement (EJ)	228	274	274	843	843	843
U_{3O8}-requirement (kton)	353	353	538	1397	1397	991
End-use share electricity	0.181	0.181	0.22	0.25	0.25	0.181

a) varying between 0.156 and 0.183 for lower and higher CO_2 emission values for methane and uranium resp. (cf. Appendix A)

The results of Table 1a show that the substitution potential, i.e., the decrease of CO_2 emission per unit of
end-use energy delivered, ranges between 2.2% and 17.7% for
the five methane-uranium substitution scenarios. It should
be noted that these scenarios are tentative; their feasibility and probability have to be assessed from in-depth mar-
ket- and technology analyses. Scenario A and B both consider a 30-year-term during which the methane end-use market
share increases to 30% at the expense of coal (scenario A)
and an increase of the electricity end-use market share is
supplied with uranium (scenario B). Only small amounts of
methane and uranium are needed over and above a no-change
scenario. A more drastic substitution over a 60-year-period
is given in scenario C, in which natural gas and electricity increase their end-use market share to 60% and 25% respectively at the expense of coal and oil products. The CO_2
substitution coefficient is 17.7 ± 2%. Cumulated demand for
methane will in scenario C will amount to about 27,000 10^9
m^3 (843 EJ), for uranium 1.4 Mton of U_3O_8. Note that the
share of renewable sources like hydropower, windpower and
biomass in the supply system has been kept constant. The
calculation for scenario D shows the impact of having to
use more (energy-) costly methane (e.g. from Siberia, the
Middle-East or deep seabeds) and uranium (average ore grade
declining from 0.1% to 0.01%). The CO_2 substitution
coefficient drops to 14.1%. Scenario E shows that a
constant end-use market share for electricity decreases the
CO_2 substitution coefficient to 15.5%, but 400 kton less
uranium-oxide is required - the equivalent of about 50
nuclear reactors of 1,000 MWe operating over this 60-year-
period.

2.3 CO_2 reduction through substitution: world

For the world as a whole, the fuel share vector has
been estimated from UN data. I have excluded wood and
other biomass and omitted bunkers (which are anyway only
half the statistical discrepancies). Then, world end-use of
hydrocarbon, hydro and nuclear resources amounted in 1986
to about 220 EJ, of which 16.3% was in the form of electricity. To estimate the share of coal, oil and gas in world
power generation, I have applied the OECD-shares. The flow
of coal, oil products and natural gas for space heating,
process heat and transport is estimated at 27 EJ, 97 EJ and
50 EJ respectively in 1986.

TABLE 1b CO_2 substitution from methane and uranium in the world

Fuel type	Fuel share F_o	Fuel share scenarios F_c				
		A	B	C	D	E
Natural gas (heat, trp) cost category I	0.225	0.45	0.6	0.3	0.6	0.525
Natural gas (heat, trp) cost category II	0	0	0	0.3	0	0
Natural gas for electricity	0.016	0.032	0.016	0.016	0	0.075
Uranium low cost 0.1%	0.025	0.05	0.2	0.066	0.115	0.054
Uranium medium cost 0.01%	0	0	0	0.066	0	0
Uranium high cost 0.004%	0	0	0	0.066	0	0
Light FOP, LPG (heat, trp)	0.385	0.285	0.06	0.06	0.11	0.24
Heavy FO (heat, trp)	0.055	0.03	0	0	0.11	0.042
Heavy FO for electricity	0.016	0.016	0	0	0	0
Coal (heat, trp)	0.012	0.02	0.06	0.06	0	0
coal for electricity	0.072	0.031	0	0	0.1	0
Other fuels (heat, trp)	0	0	0	0	0	0
Hydro and other electricity	0.034	0.034	0.034	0.034	0.034	0.034
Metallurgical coal	0.052	0.052	0.03	0.03	0.03	0.03
Anticipation period T (yr)		30	60	60	60	60
Avg annual CO_2-emission (Gton)	20.3	18.5	16.0	17.1	18.6	17.5
CO_2-substitution coefficient		0.09	0.21	0.158	0.086	0.138
CH_4-requirement (EJ)	1485	2587	5925	5925	5685	6315
U_3O_8-requirement (kton)	707	1414	8485	8485	5280	2980
End-use share electricity	0.163	0.163	0.25	0.25	0.25	0.163

Table 1b shows a number of possible scenarios for world energy supply, again keeping over-all end-use constant. Over a 30-year period, a doubling of methane and uranium shares leads to 9% lower CO_2 emission and to significant resource requirements (scenario A). Over the longer, 60-year-term, a CO_2 substitution of up to 21% seems possible in case of an increasing electricity share and a vigourous substitution of coal and oil for methane and uranium - in which case some 190 10^12 m^3 (5,932 EJ) natural gas and 8.5 Mton U_3O_8 would be required (scenario B). As scenario C shows, large penetration could require the use of lower-quality resources with a possible decrease of the substitution coefficient with 5%. A less extreme transition, but with a similar increase of the electricity share to 25%, leads to only 9% lower CO_2 emissions (scenario D).

If electricity's share is not to rise, as in scenario E, the CO_2 substitution coefficient might increase to almost 14%, showing the impact of using nuclear resources for this high-quality energy form.

3. ESTIMATING RESOURCES

3.1 Classification

Before dealing with methane and uranium resource estimates, I shall first discuss some aspects of the exploration process of which such estimates are a reflection. The real distribution of a fuel or mineral can be represented by isoconcentration curves. Within each closed curve, the average amount of fuel or mineral per unit area and for some specified depth exceeds the crustal abundance value G_{av}, which makes it an area of hydrocarbon or metal deposits. The exploration process leads to an increasing amount of experimental data, which result in combination with geological models into estimates of <u>reserves</u>, that is, known occurrences which are recoverable with currently feasible technology and at present cost. Thus: reserve estimates can not be judged outside the context of exploration history and technology cost.

On the other hand, given a set of physico-chemical characteristics like gas-bearing sediment thickness, depth and porosity or uranium ore grade distributions, one can choose boundary values of these characteristics within which a fuel or metal occurrence is classified as an "endowment" and not just common sand or rock. This, then, refers to in-situ resources and is called the <u>resource endowment</u> by Harris [4]. It is independent of technology and cost.

Several classification schemes have been proposed to deal with the uncertainty, resulting from necessarily limited data and incompletely validated models, between reserves on the one hand and resource endowment on the other hand. Here I will use the scheme of Figure 1, proposed at the 11th World Energy Conference in 1983 [5]. There are two major aspects: the geological probability of being present and the techno-economic probability of being recoverable and marketable. It has been phrased as existence vs. availability.

First the <u>probability</u> of being present. One can indicate on a geographical map areas of large exploration intensity, often expressed in length of or money expenses for exploratory drilling per unit area. Within such an area, as exploration and exploitation proceed and improvements in geology and technology develop, estimates of in-situ fuel or metal become more accurate. The usual pattern is, for a given hydrocarbon or uranium province, that additional discoveries per unit of exploratory effort decline and that the discovered occurrences can be represented by some, usually lognormal, distribution [6,7]. It adds to our estimates along the x-axis of Figure 1, filling in the domains of probable and possible reserves within the province. The dynamics of this process is some complicated function of technological, resource-economic and political variables, see [8] for the case of the oil and gas province in North-Western Europe).

Ultimate Potential Recovery					
Future Potential Recovery					Cumulative Production
Proved Reserves		Unproved Reserves		Speculative Reserves	
Developed Reserves	Undeveloped Reserves	Probable Reserves	Possible Reserves		

Figure 1 Reserve classification

For such relatively well-explored areas, one can also address the second question about recoverability. How much of a resource endowment will be technically recoverable and economically exploitable is clearly dependent on present and anticipated developments of demand, price, technology and other factors. Numerous attempts have been made to establish the impact of price change and technological developments on reserve estimates. For metal endowments, this has led to fairly well-founded, long-term supply and availability curves [9]. For natural gas, it has led attention to deep drilling, both onshore and offshore [1]. Such information allows a classification along the y-axis in Figure 1 into the domain of Future Potential Recovery, provided some set of techno- economic assumptions is given.

3.2 Estimating the unknown

However, the more critical question in long-term resource appraisal is how to assess reserves in less well-explored or unexplored areas. This is an important issue: in 1981, 73% of all 3.78 million wells drilled in the world were in the U.S., 90% of all exploratory drilling in the world between 1971 and 1981 has been in the industrialized countries. For uranium, too, the reserves per unit area are at present much larger in a few well-explored countries than elsewhere. The "art" of estimating reserves in the vast unexplored areas of the world is one of inference.

Several methods have been developed for such, necessarily speculative, estimates of the unknown. Broadly speaking, they can be grouped into three approaches:
a. statistical analysis of past trends in exploration and exploitation
b. statistical analysis of characteristics of deposits thus far discovered
c. subjective and/or objective probabilistic methods based on geological expertise.
King Hubbert's (1969) well-known analysis [10], based on the declining rate of discovery of oil and gas in the U.S. for increasing cumulative exploratory footage is an example of the first method. Lieberman [11] has used the same approach to US$8/lb U_3O_8 endowments. Another approach is to extrapolate average fuel or metal found per km^2 of sedimentary basin, per km^3 of rock etc. The third approach diverges into a rather wide spectre, including Delphi methods and probabilistic models (see e.g., [6,12]). Of course. The three approaches are not strictly separate and sometimes are merged on purpose.

Characteristic for all three approaches is the use of some relationship of analogy or similarity between well-explored and unexplored areas. However, the nature of this relationship differs. Statistical trend analysis presumes not only geological but also technical and economic similarity, whereas spatial distribution models infer the unknown from some underlying over-all statistical distribution of fuels and metals. The probabilistic methods rely on expert knowledge on specific fossil fuel or mineral provinces, similarity usually being defined in geological terms (see e.g. [6] for a formal similarity matrix for petroleum).

Using one or more of these methods one may estimate a hydrocarbon or metal endowment in a country, a continent or - most audacious - in the earth's crust. Such estimates are often referred to as ultimate resource-in-place. Evidently, statements about the availability of these endowments in place and time are necessarily highly speculative. In 1900, uranium was no energy resource at all. In 1970, no one included geopressurized methane into mankinds endowment.

3.3 Cut-off (energy) costs

An interesting and decisive factor in estimating resource availability is the choice of cut-off value for one or more of the physico-chemical characteristics or some function of them, beyond which known or unknown occurrences are no longer considered available for use. Money costs are the most common cut-off variable. Usually, it is based on some engineering cost function, which relates resource cost to e.g. deposit size, grade and location [8]. An obvious though often neglected problem with such a measure is, apart from the economic and technological uncertainties involved, that it is impossible to say which level of marginal resource costs are realistic from a societal point of view. Does it make sense to assess uranium resources up to ten times the present marginal cost, assuming that nuclear power will still be competitive at such cost levels? Or, as I would argue, do such resource estimates only make sense in the context of an explicit long-term scenario for world economic development?

A second measure is based on net energy analysis (cf. [39]. In this view, there is a hard and important measure for the maximum allowable energy resource quality decrease, namely the ratio between end-use energy delivered and energy inputs all along the way from exploration to end-use delivery. If this so-called energy ratio falls down to levels close to or even below one, there is no longer any net gain and the system is evidently unsustainable at a global long-term. This measure also has to deal with uncertainties about prices and technology. However, in the present context the choice of a maximum allowable energy ratio appears to be less arbitrary than in case of money costs: evidently, the indirect CO_2 emissions should not become so large as to offset the gains in CO_2 emission reduction from lower direct CO_2 emissions. I will subsequently return to this point later on.

4. METHANE

4.1 Supply from proved and unproved reserves

Two characteristics have dominated the market penetration of natural gas, of which methane is the most important constituent: its association with oil and the requirement of a capital-intensive infrastructure. Natural gas has been closely associated with oil since the beginning of the petroleum era, first to drive oil out or be flared, later to become also a highly valued fuel. Whiting [6] estimated that in 1975 40% of gas reserves was such associated gas, found for over 80% in less than 10 large oil fields. The other 60% of gas reserves, containing so-called non-associated gas, are for two-third concentrated in 119 fields of over 1 EJ each. Thus, there is evidence of a lognormal distribution, which points to the importance of unexplored areas for major additions to proven reserves.

For associated gas, of which in 1980 55% of production was still flared [13], improved petroleum reservoir engineering may increase the gas supply potential. An interesting trend is the relative increase of non-associated gas reserves. Some geologists argue in favour of large, deep (and possibly abiogenic) methane occurrences, for which the conventional oil-based estimation methods fail. The rapid increase in exploratory drilling at depths below 4000 m. might, in this view, explain why the fraction of non-associated gas in rising [2]. Yet, we shall first look at the less speculative reserves.

Figure 2 shows the development of cumulated production and the proved and unproved reserves between 1970 and 1988. Despite a 50% higher usage rate, the ratio between proved reserve and annual production for methane worldwide has improved. Moreover, there is a fair consensus that as yet undiscovered reserves will be at least as large as the proved reserves. A recent example of an assessment of worldwide known and undiscovered potential reserves of gas is given by Colliti [14]. Based on statistical analysis of field size distribution and exploration intensity and inference onto less explored areas, he expects that some 135 10^{12} m^3 (4,300 EJ) of gas reserves can be discovered and extracted under the techno-economic conditions of the 1990s. To ascertain this amount of 6,600 EJ (including 1979 known reserves) requires large exploration efforts.

As can be seen from Table 1, a total reserve of 10,000 EJ would satisfy world demand until 2050 in case of a strong further penetration of methane into a non-growing end-use energy market. Important questions are: can the market absorb gas at the rate of the scenarios in Table 1 and will there be a timely and reliable supply capability? Odell [15] has expressed the view, that the European market may become the favourite target of export-ambitious suppliers like the USSR and Iraq. This could raise the gas supply potential to as much as 14 EJ/yr by 2000 - 44% of end-use in EUR-12 in 1987 (cf., Table 1). For the world as a whole, the World Energy Conference [16] estimated natural gas production capability at 145 EJ/yr in 2000, with large regional differences in growth rate. Assuming end-use stabilisation this would imply a penetration rate of over 60%.

However, gas supply capability does not easily match demand from a geographical point of view. More than 70% of proved and additional reserves of 9,700 EJ as estimated by the International Gas Union in 1986 are located in the USSR, North Africa and the Middle East [17]. It means that the large European and Asian markets can only be served with a significant increase in international methane trade

either by pipeline or LNG-carrier. In 1980 imports and ex-
ports accounted for less than 10% of consumption [38]. It
remains to be seen whether the international financial and
political system will be able to support an increase in
methane trade towards the size and complexity of present
crude oil import-export patterns.

4.2 Speculative reserves

In view of justified doubts about the financial and
political feasibility of large-scale methane penetration
based on gas reserves in well-explored provinces, it is
legitimate to investigate the potential of more speculative
methane reserves. This is the more relevant because the
substitution scenarios of Table 1 presume a gradual transition towards new energy supply options after 2050. This
point is highlighted in Marchetti's approach [1]. In the
logistic substitution model, it is suggested that we are at
the beginning of the "methane age": gas demand increases to
a share of more than 55% worldwide, with a peak of 950
EJ/yr around 2060 and a cumulated life-cycle demand of
80,000 EJ in the period 1900-2150. This is ten times the
present proved and unproved reserves - can we count on such
large quantities with anything more than speculation and
faith? But also the more prudent development of constant
end-use, as sketched in Table 1, would require some 12-
15,000 EJ of methane over the whole life-cycle and even
more if nuclear energy would be an unacceptable large-scale
alternative.

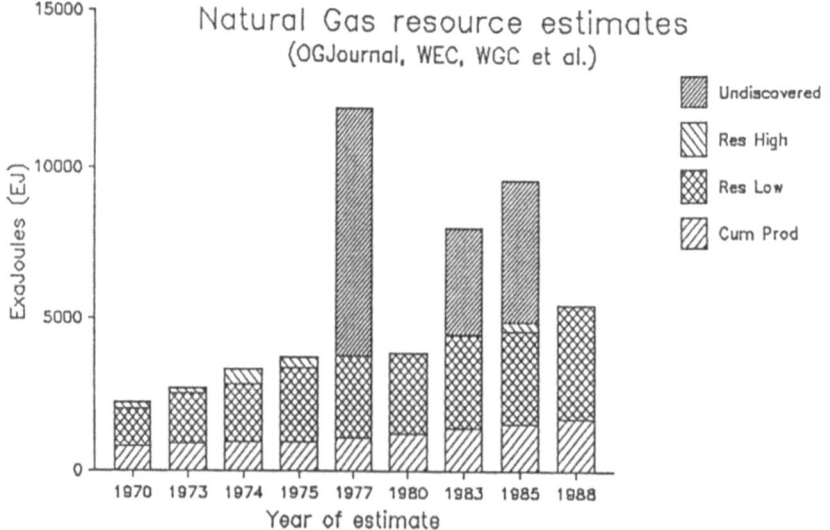

Figure 2 Methane reserves

Unfortunately, uncertainty dominates the geological as well as the techno-economic aspects of the speculative or "unconventional" gas reserves. There have been speculations in the last decades about the existence of giant amounts of natural gas in deep and geopressurized zones. Table 2 gives some estimates of these reserves, derived from geological analogy. Most of them come in through the broadening of characteristics like porosity, permeability, density, depth and the like.

A first group are the tight sandstone formations: large volumes of sandstone and/or shale with 5-15% porosity, large water saturation and low gas permeability. To extract the gas out of it, several techniques are being developed of which massive hydraulic fracturing of the reservoir is the most promising. At present, gas recovery from tight formations in the US amounts to about 1 EJ/yr. Large-scale exploitation requires enormous amounts of fluids and sand, with adverse environmental impact; technical and economic recoverability are still uncertain.

Another other major speculative reserve is geopressurized gas. As with the speculation about abiogenic methane at great depth, scientific controversy abounds. Dorfman [18], in his optimistic assessment, suggested 0.2 EJ/yr by 2000 as a timetable for development in the US. Technical and economic prospects for the recovery of dissolved natural gas from geopressured zones are uncertain; enormous amounts of water have to be produced, with unknown environmental impact. Perhaps one of the most serious barriers to the exploitation of geopressurized gas is the low net-energy gain of this diffuse source. Hall et al. [7] have evaluated the energy ratio of geopressurized gas from the US Gulf Coast region. The most promising wells had produced 1 to 5.5 times more energy than the energy input required. This suggests that even vast geopressurized zones may have a small <u>net</u> methane production potential.

A third, more recently discussed potential methane reserve are clathrates [21]. These are crystalline compounds in which methane is trapped in expanded ice lattices in densities which are up to 50 times higher than in shales, tight formations and geopressirized zones. According to MacDonald [21], methane can be recovered from Arctic permafrost regions at a positive energy ratio of 13 i.e. some 92 % efficiency, using thermal stimulation. Methane from the Alaska North Slope might be available in the Pacific region at similar costs as from other unconventional sources. Based on geochemical data and exploratory drilling experience, MacDonald derives a reserve estimate of an astonishing 21.000 10^{12} m^3 - five times estimated coal reserves. Commercial development is most likely to start in the Arctic regions, but vast technological and logistic problems have to be overcome before any sizeable contribution can be expected. It is noteworthy to recall, that the large permafrost methane occurrences have also been discussed as a possible positive feedback element in the greenhouse dynamics.

5. URANIUM

5.1 Supply from assured and additional reserves

In the decade after the first "oil crisis" in 1973 there has been great interest in uranium reserves and availability, due to an anticipated enormous growth in nuclear power capacity. Uranium exploration expenditure in non-communist countries has, however, drastically fallen down from a high $ 700 million in 1980 to about $ 150 million in 1988 [22]. This reflects, in the aftermath of the Harrisburg and Chernobyl accidents and successful energy conservation in a slow-growth economy, the greatly reduced expected demand for uranium. Whereas the World Energy Conference [23] estimate of cumulated U-demand for the period 1975-2020 varied between 6 and 10 Mton (mln ton), it is down to less than 0.5 Mton until 2000 in more recent estimates [24]. Consequently, not much research has been done to increase our knowledge of uranium availability.

TABLE 2. Speculative gas reserves

Type	Estimate (EJ)	Source
Unspecified USA	895-1,345	Hefner [19]
Unspecified N-Am.	1,790-2,690	Hefner [19]
Deep source & hydrates N-Am.	5,125	Hanneman [20]
Coal-bed degasification USA	325-870	WEC [16]
Devonian shale USA	545-650	WEC [16]
Tight formation USA	650	WEC [16]
Geopressured gas USA	3,200-54,400	WEC [16]
Tight sandst form USA	175	Meyer [6]
	350-665	Meyer [6]
Geopress zones N-Am.(3-7 km)	2,700-44,000	Dorfman [18]
Shale and tight sandst USA	1,075	ERDA-1977 [21]
Coal seams	2,200	ERDA-1977 [21]
Geopress aquifers USA	2,690-88,600	ERDA-1977 [21]
Methane Clathrate Siberia	24,000	McDonald [21]
Methane Clathrate Ocean sed	635,000	McDonald [21]

Figure 3 shows reserve development in Mton U_3O_8 over the last 15 years. Cumulated production is approaching 1 Mton. The conventional classification in "reasonably assured" and "estimated additional" is roughly parallel to "proved" and "unproved" reserves (Figure 1). The cost categories I and II are indicated in current dollars. Usually, costs in $/kg U is related to average ore grade (in U_3O_8 weight percentage). Category I are the rich ores of more than 0.1 wt % U_3O_8. Category II refers mostly to ore grades between 0.1 and 0.01 wt % U_3O_8 but the boundaries are neither strict nor fixed. The estimates in Figure 3 indicate, that relatively low-cost uranium reserves have consistently be estimated in the range of 5-7 Mton U_3O_8. Looking at Table 1a-b, one may conclude that any vigourous transition towards uranium, as for instance suggested in the IIASA logistic substitution forward calculation [2], would deplete these reserves within the next 60 years with present nuclear technology. Moreover, the assumption that the transition away from uranium to some other future energy source should be at the same rate as its introduction implies that presently estimated reserves seriously fall short. Due to the large financial risks involved in large-scale expansion of the capital-intensive nuclear cycle, in the meantime a shortage in uranium production capability could occur (cf.,[25].

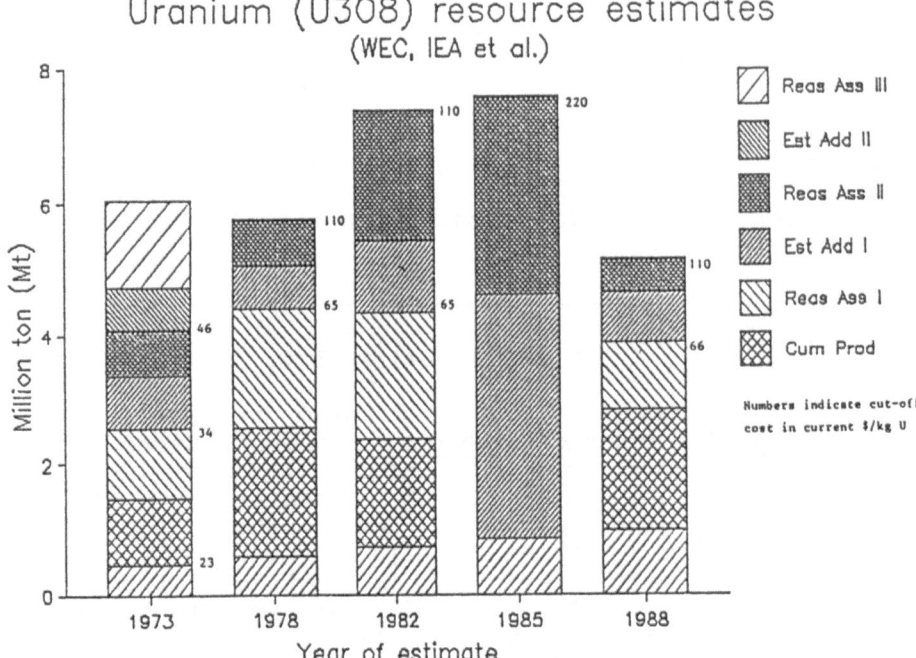

Figure 3 Uranium reserves

5.2 Speculative reserves

An important question, then, is whether there are more low-cost speculative reserves. The 1978 WEC analysis has discussed these in some detail; more recent evaluations are given by Harris [4]. Harris presents a nice illustration of how speculative reserve estimate is dependent on methodology. Method A to infer world known and unknown uranium reserves is based on transferring US data on reserve development onto the rest of the world (historical analogy). Method B applies the proposition that for large enough regions there should be an equal amount of uranium per unit of area or volume all over the world (geographical analogy). The third method, C, combines analysis of exploration, geographic analogy and economic adjustment. Table 3 shows Harris' as well as some other estimates of low-cost proved, unproved and speculative reserves (including cumulated production). One first should note the amount: two to ten times assured and additional reserves. Secondly, the three methods yield widely varying estimates - apparently mankind is not able to make a more reliable assessment: "At our current level of information, it is impossible to know what number to use" [4, p. 111].

TABLE 3. Uranium reserve estimates in Mton
(incl. cum. prod.)

	< US$ 66 (1978)/kg U_3O_8	< US$ 110 (1978)/kg U_3O_8
Method A	10	14
Method B	19	25
Method C	9-22	12-29
Brinck [28]	80(<13(1971)$)	300(<18(1971)$)
Cameron [29][a]	<$13 ~ 12	<$ 18
IEA [30][b]	~ 24	

[a] "ultimate"
[b] "speculative", < US$ 220 (1984)/kg U_3O_8

A next step in assessing future uranium availability focuses on known and unknown high-cost reserves. Based on geological exploration, one may infer very large low-grade (<0.01 wt % U_3O_8) reserves in e.g., granitic rock and shales. Converting grade into cost, some authors have attempted to construct a long-term supply curve. For example, Grenon [12] estimated 10 Mton to be available at costs below US$ 220 (1975)/kg U_3O_8 but 40 Mton (>0.001 wt %) below 400 and 4,500 Mton (seawater) below US$ 660 (1975)/kg U_3O_8. Indeed, these large amounts have high probability-but how about recoverability?

5.3 The Crustal Abundance Geostatistical (CAG) approach

One of the formal models which look at uranium reserves from such a global and long-term perspective, is the Crustal Abundance Geostatistical (CAG) model. Originally formulated by De Wijs and Brinck, it is based on the empirical fact of logbinomial element distribution in the earth's crust (for details, see De Vries [26,8]). Given an estimate of the average uranium concentration or crustal abundance G_{av} and the so-called specific mineralisability Q which is a measure of the metals tendency for dispersal through natural processes over the last billions of years, one can calculate the expected U-reserves as a function of cut-off grade G_{co}.

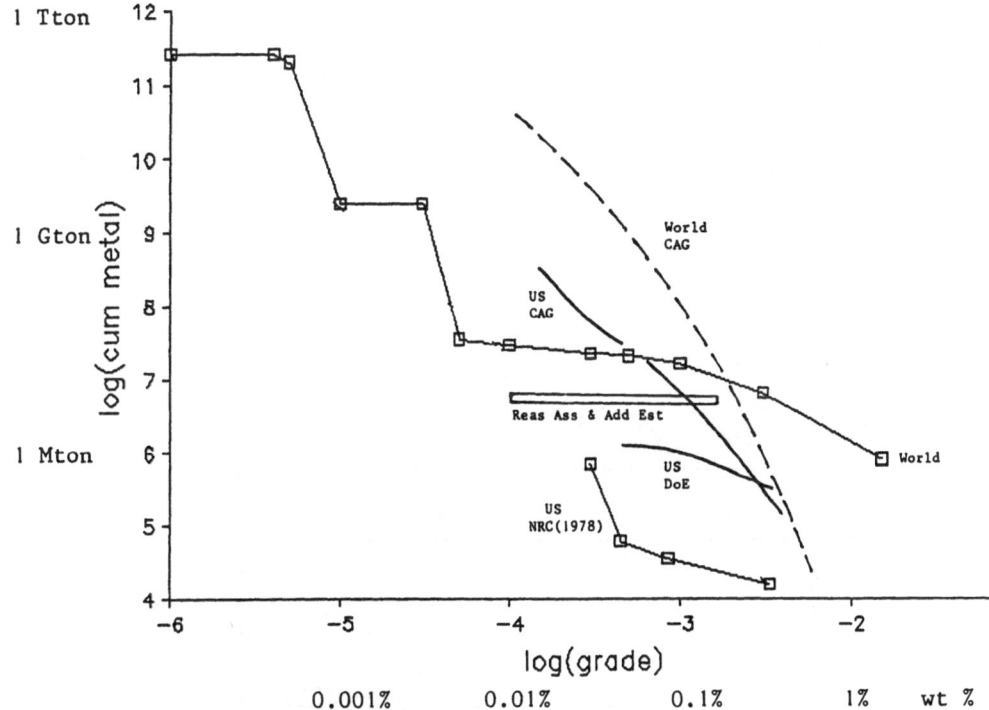

Figure 4 Cumulated uranium reserves vs grade

Figure 4 shows the reference calculation (G_{av} = 2.1 ppm, Q = 0.2) together with some independent estimates of cumulated reserves vs. grade. The three lower curves show cumulated reserves vs. grade for the US. The CAG-approach tends to underestimate high-grade reserves, but infers large amounts of lower grade deposits. As Harris [4]

232

comments: "... the large estimates of endowment that are produced by CAG-models have been a barrier to the acceptance by many geologists of these models". The block-shaped curve is based on geological estimates of separate occurrences [27] - evidently, the CAG-approach suggests larger reserves below 0.1 wt % U_3O_8 than geological exploration up to the present day indicates.
Brinck [28] has proposed to construct long-term availability curves of cumulated metal vs. marginal cost through the use of engineering cost relations. The dominant features are mining cost decreasing with deposit size, and milling cost decreasing with grade. In the present context, as I have explained in paragraph 3.2, it is meaningful to analyse the resource endowment in economic and also in energy costs.

5.3.1 Economic cut-off cost. We shall assume that an engineering cost function relates cost per ton of U_3O_8 to deposit size, grade and some other factors (see Appendix B). Now, one can calculate so-called equal-cost U-occurrences which are somewhere in-between small, high-grade and large, low-grade. Such isocost-curves are shown in Figure 5a for

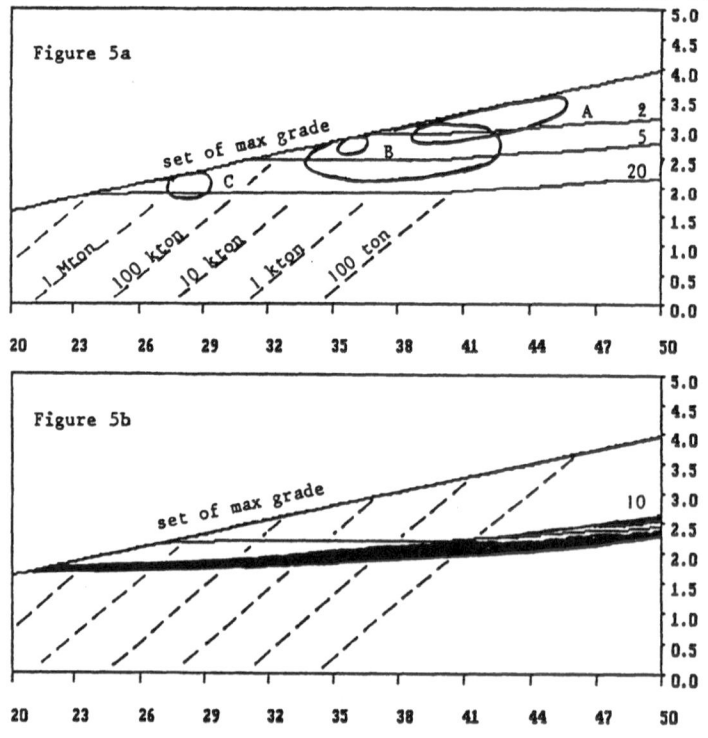

X-axis b with Deposit Size=10**18/2**b ton, Y-axis log(Gdep/Gav)

Figure 5 Isocost curves for uranium

three values of C/C_{ref} with C_{ref} the cost of some known reference deposit. Along the x-axis is the 'grid parameter' b, indicating how coarse our geographical division of the earth's crust in equal-sized blocks is. It is a measure of deposit size $S = M/2^b$ with M earth crust mass. Along the y-axis I have plotted the logarithm of the natural concentration factor G/G_a v with G the average grade of the block of corresponding b-value.

I have also drawn isometal-curves as straight, dotted lines. The figure shows that a given amount of U_3O_8 can be found in small blocks of high average grade (high b) or large blocks of low average grade (small b). Going downward along an isometal-curve, grade will decrease and cost and probability will increase. I have drawn the tonnage-grade domain for a number of representative uranium occurrences: ore veins (A), sandstone type and calcrete (B) and Swedish shale (C).

Figure 5a indicates that at 20 times higher cost than the rich reference deposit, one may expect even Swedish shale to be an economic resource - indeed, these shales contain 77% of the 1978 estimate of reasonably assured U-resources in Western Europe [23]. It also suggests at this cost level a minimum cut-off grade of about 0.016 wt % U_3O_8 and cumulated metal reserves in the order of 10 Mton. The isocost-curves slope upward for smaller blocks (larger b) because there are much larger scale economies below 1 Mton block size. The impact of this assumption is shown in Figure 5b for 10 times higher than the reference cost. The black band indicates that with the same economies of scale for large as for small deposits, the lowest cut-off grade would drop from 0.03 wt % to below 0.01 wt % U_3O_8. 'Economic reserves' at this cost level of about US$ 100(1975)/kg U_3O_8 would increase by a factor of ten! This shows the importance of such a rather neutral assumption on the economies and technology of scale.

5.3.2 **Energy cut-off cost.** Analogous to cost relationships, expressions for energy required per ton U_3O_8 have been formulated (see Appendix B). Thus, the CAG-model enables us to relate reserve estimates to the net CO_2 substitution coefficient of uranium by drawing iso-energy-cost curves. Figure 6a shows these curves for energy costs of 10, 20, 50 and 100 times the 0.56 TJ required to get one ton U_3O_8 from the rich reference deposit. Most energy analyses agree that nuclear energy is no longer a net CO_2 substitution option with regard to coal/oil and methane respectively for energy cost being 100-150 and 50-100 respectively times the energy cost of the rich reference U_3O_8-deposit.

234

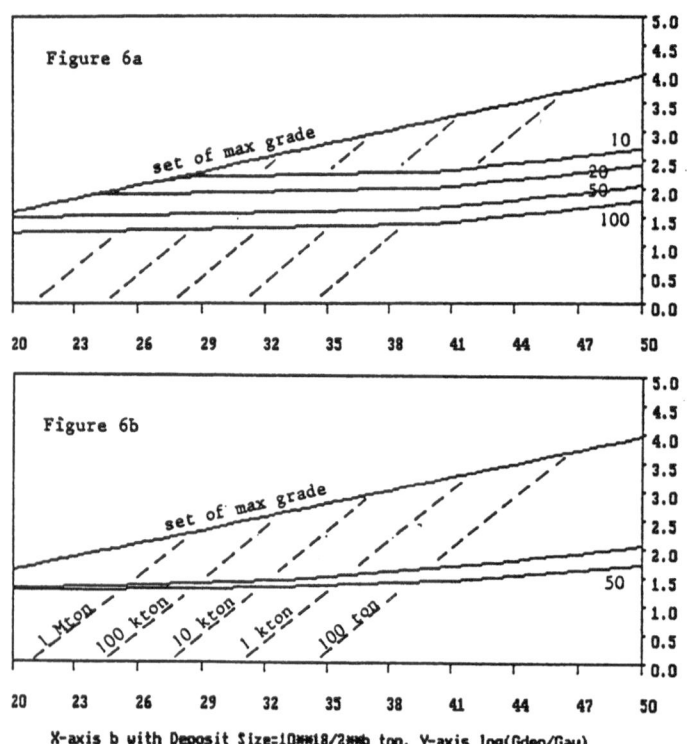

X-axis b with Deposit Size=10**18/2**b ton, Y-axis log(Gdep/Gav)

Figure 6 Iso-energy-cost curves for uranium

Thus, the curves in Figure 6a suggest that the first 10 Mton will certainly contribute to CO_2 reduction if it replaces coal. As Figure 6b shows, the assumption of continuing (instead of decreasing) energy economies of scale would bring this point of zero net CO_2 reduction virtually to infinity. On the other hand, the tremendous environmental problems in mining low-grade ores might as well imply a negative energy economy of scale - in which case the lean ores among the reasonably assured reserves could already have a negative net CO_2 reduction in substituting uranium for natural gas! Here, again more analysis of the plausibility of engineering (energy) cost relationships is needed to assess uranium in LWR as a net CO_2 substitution option for the long term.

5.4 Some comments

What is to be concluded from these results? In my view, the CAG-model provides a coherent framework for uranium exploration and exploitation, inferring unknown reserves from the widely accepted hypothesis of lognormal element distribution. This is illustrated with figure 5a. Exploration usually proceeds in such a way, that some rather rich deposits are discovered and exploited first (e.g., A). Then, larger leaner deposits (e.g., C) are discovered or, if already known, classified as reserves and taken into exploitation while developing the necessary technology. This explains why large reserve additions of a given cost category show up after the depletion of the rich provinces. However, if there are still large unexplored regions, one may expect new discoveries of probably high-grade ore (e.g., B). It should be noted here that

- it is not assumed that reserves are exploited in order of decreasing ore grade (although average grade of processed ore in the US has already dropped from over 0.2 wt % to less than 0.15 wt % U_3O_8), and
- part of what has been attributed to cost-lowering technological progress (cf. [31]) or a learning curve (cf. [32]) is actually included here insofar as a transition to larger deposits yields economies of scale (the coefficient τ).

The two major objections against the CAG-approach are that the model parameters are hard to estimate empirically and that some geologists have hypothesized the existence of a bimodal element distribution and/or a discontinuous change to rock silicate occurrence (the "mineralogical barrier") due to which extraction costs jump abruptly. As to the latter: "no one has demonstrated by data the existence of a bimodal distribution, but, it is likely that no one yet has had sufficient data to perform a credible test of this hypothesis" [4, p. 231]. As to the empirical parameter estimate, Holland and Peterson [41] warn against too much reliance on CAG-endowment estimates but provide no alternative: "A reasonably accurate figure for total resources will have to await the results of another century of exploration".

The geostatistical approach provides an appropriate way to analyse and discuss some of the crucial aspects of long-term uranium resource appraisal. A more rigourous inclusion of past exploration history and results could improve its empirical basis; more robust (energy) cost relations based on technology trends would allow a more reliable supply cost curve. In the meantime, we have to accept that there simply is no well-founded answer to how much uranium can be supplied at what cost, beyond the estimate of 5-7 Mton U_3O_8 from well-explored regions.

6. CONCLUSIONS

Penetration of methane and uranium into the energy market to fractions of 50-60% respectively 10-20% of present end-use can lower cumulated CO_2 emissions with some 15-20%. If end-use can be stabilized at the present level, this would require world-wide some 6,000 EJ of natural gas and 5-8 Mton U_3O_8 between 1990 and 2050. A smooth transition to another energy source afterwards could double these requirements, as would the widely used assumption/goal that energy end-use should double over the next decades.

Present estimates of proved and unproved methane indicate no major resource constraints for end-use stabilisation scenarios. However, if cumulated methane requirements should exceed 10,000 EJ, mankind will probably have to exploit yet speculative reserves at higher costs and lower net energy gain. This, in turn, could reduce the net CO_2 substitution potential of substitution with several per cent points. Estimates of methane production capability suggest the technical feasibility of rapid, large-scale penetration of methane. However, a serious impediment may be financial and political, in view of the massive infrastructural capital investments required and the geographical concentration of proved and unproved gas in only a few provinces.

The reasonably assured and estimated additional reserves of U_3O_8 are sufficient to meet worldwide penetration of nuclear energy if end-use stabilisation is assumed. This could be at cost below two-three times present costs. More advanced nuclear technology will improve the prospects; more cautious, higher estimates of mining and milling costs for low-grade U_3O_8-reserves may significantly reduce the net CO_2 reduction potential of the nuclear option. The geostatistical model highlights the fact that the amount of low-grade reserves is very large indeed, but that any estimate of the economic and energy costs at which they can be exploited, is based on inference from present mining and milling experience with high-grade deposits. One is left, then, with the certainty of large reserves but without reliable indication of their (energy) cost level and an appropriate (energy) cut-off point.

In summary, I conclude that methane and uranium can play a significant role in reducing CO_2 emissions if energy end-use can be stabilised at the present level. The prospects for methane are good, although the vast capital outlays for transport and distribution and the required international agreements may make rapid penetration illusory or even dangerous. For nuclear energy, equally large capital investments are needed to have uranium play a role in CO_2 reduction. Strategic aspects seem less important than the anticipated decrease in net CO_2 substitution once lower-grade ores have to be exploited.

Combined with the economic growth to which developed and developing nations are equally committed, end-use stabilisation will require vast efforts in such fields as energy conservation and renewable energy sources. If these options can not compete successfully with the more concentrated and powerful demand for supply-oriented capital, any CO_2 reduction strategy may be thwarted : net CO_2 reduction may be unattainable, speculative gas reserves have to be counted on, leaner uranium occurrences will have to be exploited sooner. The real challlenge ahead lies in a balanced combination of demand and supply options.

Appendix A. CO_2 emission values

Data for direct and indirect CO_2 emission values have been based on the literature [2,33,34,35,7]. Standard conversion factors from Joule to physical units are also given and included for electricity average end-use efficiency.

Indirect CO_2 emission coefficients are based on various assumptions on energy quality and fuel use. The indicated uncertainty in the estimates reflects technological factors (e.g. electricity conversion efficiency and LWR fuel cycle options) as well as the differences in direct CO_2 emission estimates.

TABLE Direct and indirect CO_2 emission coefficients (kg/GJ)

	direct		indirect		conversion	
	EUR-12	World	Eur-12	World	factor	
natural gas Heat/Trp	55±1	53±2	1	2	31.65	MJ/m3
natural gas Heat/Trp	55±1	53±2	7±2	7±2	31.65	MJ/m3
nat.gas Elec(44%E 5%L)	125±3	130±3	10±6	10±6	13.93	MJ/m3
uranium 0.1%(34%E 5%L)	0	0	21±2	21±2	175.00	TJth/tonU$_{3O8}$
uranium 0.01%	0	0	40±6	40±6	17.50	TJth/tonU$_{3O8}$
uranium 10-20 ppm	0	0	75±15	75±15	7.00	TJth/tonU$_{3O8}$
light FOP/LPG Heat Trp	69±1	69±2	10±2	11±2	45.00	MJ/kg
heavy FO Heat/Trp	80±5	80±5	8±2	9±2	45.00	MJ/kg
heavy FO Elec(40%E 5%L)	200±5	200±5	22±3	26±3	18.00	MJ/kg
coal Heat	90±2	90±4	7±1	10±2	26.90	GJ/ton
Coal elec (38%E 5%L)	245±5	245±8	25±5	35±10	10.30	GJ/ton
other non-CO_2 Heat/Trp	96±1	96±1	1	1		
other non-CO_2 Elec	0	0	3	3		
coal metallurgical	101	101	6	6	30.00	GJ/ton

Appendix B. Cost engineering relationships for uranium

Economic cost of uranium in the form of U_3O_8 contains several terms: mining cost, milling cost, transport cost and an aggregate cost term for exploration and development drilling. Mining costs are mainly dependent on deposit size and depth and, relatedly, mining method (surface/underground). Milling costs, upgrading the ore into high-grade U_3O_8, are often taken proportional to ore grade per ton of ore, that is, inversely proportional to ore grade per ton of U_3O_8. The general relationship is

$$C \text{ (\$/ton } U_3O_8) = c_1 (S/S_{ref})^\sigma/G + C_2 (G/G_{ref})^\tau$$

in which ref refers to some reference deposit, S to average deposit size and G to average deposit grade. A representative rich reference deposit is 1 Mton rock of 0.4 wt %, containing 4 kton U_3O_8. All cost can be expressed in the cost of such a reference deposit, C_{ref}. Common values for σ are -0.1 to -0.3 and for τ in the order of -1 [26].

For the present purpose I have used a relationship discussed by [4] for the New Mexican uranium endowment. Mining cost are given as 21.5 $(S/S_{ref})^\sigma$ \$/ton ore, with σ = -0.11 for S<1 Mton and σ = -0.01 for S>1 Mton. Milling costs are estimated at 6.4 + 559.7 G \$/ton ore. Other costs are estimated about \$ 3,000/ton U_3O_8. The resulting cost equation is:

$$C=3000 + 6.4/(nG) + 559.7/n + 21.5(2^\sigma(b_{ref}-b))/(nG)$$

in US\$(1975)/ton U_3O_8 with the extraction conversion calculated according to n = 0.98 - 0.0723 [log(100 G)]2 [22]. For the reference deposit this yields about 10.3 \$/kg U_3O_8 or 12.2 \$/kg U. Note that S has been converted to the CAG-parameter b through the relationship S= $10^{18}/2b$ (cf. [26]).

Energy cost or the GER (Gross Energy Requirement) of U_3O_8 have been approximated with similar relationships. Chapman and Hemming [36] give a loglinear relationship of the form GER=α G^β TJ/ton U_3O_8 with α= 0.0013, β= -0.98 for no overburden and α= 0.004, β = -1.2 for a stripping ratio of 25. Smith [22] uses GER = $\alpha/(nG)$ TJ/ton U_3O_8 with α ranging from 0.0023 (USA) to 0.0126 (South-Africa). Horsten [37] uses earlier US-estimates: for mining GER = 0.01 $G^{-0.65}$ and for milling GER = 0.001 $G^{-1.025}$ TJ/ton U_3O_8. Remarkably, none of these authors has assumed energy economies of scale. In the present report I use the following relation:

$$GER = e_1 (S/S_{ref})_{\sigma}/G + e_2 G^\tau \text{ TJ/ton } U_3O_8$$

in which the rising energy costs for mining are attributed to both decreasing grade and deposit size. From the previously cited references and to ensure convergence, I have derived the parameter estimates: $e_1 = 0.0011$, $\sigma = -0.2$ (but one-fourth of this value for deposit size > 1 Mton), $e_2 = 0.001$ and $\tau = -1$.

LITERATURE

1. Lee, T., H. Linden, D. Dreyfus and T. Vasko (eds) (1988). The Methane Age, Kluwer/IIASA.
2. Ausubel, J., A. Grübler and N. Nakicenovic (1988) 'Carbon Dioxide Emissions in a Methane Economy', IIASA RR-88-7.
3. Kram, T. and P. Okken (1989) 'CH_4/CO-emission from fossil fuel global warming potential', ESC-WR-89/12.
4. Harris, DeVerle P. (1984).Mineral Resource Appraisal, Oxford: Clarendon Press.
6. Meyer, R.F. (ed.,) (1977).The Future Supply of Nature-Made Petroleum and Gas. Technical Reports', Pergamon Press.
7. Hall, C.A., C.J. Cleveland and R. Kaufman (1986). 'Energy and Resource Quality', Wiley-Interscience.
8. Vries, B. de (1989). 'Sustainable resource use - an enquiry into modelling and planning', Ph.D. Dissertation Groningen.
9. BuMi (1983,1984) - U.S. Bureau of Mines Information Circular 1983-8930 (Copper Availability) and Cicular 1984-8978 (Manganese Availability).
10. Hubbert, M.K. (1969).'Energy resources', in Resources and Man, National Academy of Sciences, W.H. Freeman, San Francisco.
11. Lieberman, M.A. (1976).'United States uranium resources - an analysis of historical data', Science 192 (4238) p. 431-436.
12. Grenon, M. (ed.,) (1979).'Methods and Models for Assessing Energy Resources', IIASA Proceedings Series Vol. 5, Oxford: Pergamon Press.
13. Wionczek, M.S. (ed.) (1983). World Hydrocarbon Markets, Oxford: Pergamon Press.
14. Colliti, M. (1983). 'Size and distribution of known and undiscovered petroleum resources in the world with an estimate of future exploration', in World Hydrocarbon Markets, M.S. Wionczek (ed.,), Oxford: Pergamon Press.
15. Odell, P.R. (1987).'Prospects for West-European energy markets', Energy Papers 14, Foreign Policy Institute, John Hopkins University.
16. WEC (World Energy Conference) (1978). 'Oil and Gas Resources', IPC Science and Technology Press.

17. Frisch, J.R. (1987). 'Future stresses for energy resources', Graham & Trotham.

18. Dorfman, M. (1976). 'The supply of natural gas from geopressured zones: engineering and cost', in The Future Supply of Nature-Made Petroleum and Gas: Technical Reports, R.F. Meijer (ed.,), Pergamon Press.

19. Hefner, R. (1988) 'Energy: economic and geostrategic considerations', in Kluwer/IIASA The Methane Age, T. Lee et al., (eds.).

20. Hanneman, R.E. (1988). 'Methane technology: a technical survey', in Kluwer/IIASA The Methane Age T. Lee et al., (eds.,).

21. McDonald, G.J. (1989). 'The near- and far-term technologies, uses, and future of natural gas', Paper presented at the IEA/OECD Expert Seminar on "Energy Technologies for Reducing Emissions of Greenhouse Gases", Paris.

22. Smith, P.B. (1982). 'Energietoevoer van de LWR 'once through' cyclus vergeleken met de geproduceerde energie', Bijlage E in J.W. Storm van Leeuwen (1980) 'Energie-analyse van een PWR kerncentrale', CE Delft.

23. WEC (World Energy Conference) (1978). 'Nuclear Resources', IPC Science and Technology Press.

24. ATW (Atomwirtschaft) (1988). 'Die Uranversorgung der Welt', ATW 12(1988) 600.

25. Fujita, A., and P. Silvennoinen (1985). 'Uranium resource appraisal', Resources Policy, June 1985.

26. Vries, B. de (1988). 'Sustainable resource use - optimal depletion within a geostatistical framework', IVEM-35 RU Groningen.

27. Bijlsma, J., K. Blok en W. Turkenburg (1989). 'Kernenergie en het kooldioxideprobleem', N&S RU Utrecht.

28. Brinck, J.W. (1979). 'Uranium resourcxes assessment with "MIMIC": a descriptive model of mineral resources and long-term price trends in the mining industry', in Methods and Models for Assessing Energy Resources, M. Grenon (ed.,) IIASA Proceedings Series Vol. 5, Oxford: Pergamon Press.

29. Cameron, E. (1979), in [12].

30. IEA (1983), in [27].

31. Barnett, H.J. (1979). Scarcity and Growth revisited' in V.K. Smith (ed.) 'Scarcity and Growth Reconsidered: Resources for the Future, Baltimore: The John Hopkins University Press.

32. Chapman, P., and F. Roberts (1983). 'Metal resources and energy', Butterworths London.

33. Blok, K., S. Fockens, J. Bijlsma and P. Okken (1988). 'CO_2 Emissiefactoren voor brandstoffen in Nederland', N&S RU Utrecht/ESC Petten Rapport ESC-WR-88-12.

34. Krause, F. et al. (1988).'Energy and climate change: What can Western Europe do?', IPSEP-Report. Richmond

242

CA 94805.

35. Vries, B. de and E. Nieuwlaar (1981).'Coal for electricity in the Netherlands - energy, environment and cost aspects', IVEM RU Groningen.

36. Chapman, P. and D. Hemming (1976). 'Energy requirement of some energy sources', in the 9th International TNO-Conference 'The Energy Accounting of Materials, Products, Processes and Services'.

37. Horsten, J. (1982).'De energiehuishouding van een lichtwaterreactor', TH Eindhoven, afd. Technische Natuurkunde.

38. Wionczek, M.S. and M. Serrato (1983)'The present and future of natural gas', in World Hydrocarbon Markets, Oxford: Pergamon Press.

39. Nieuwlaar, E. (1988). 'Developments in Energy Analysis', Ph.D. Dissertation Utrecht.

40. NRC (National Research Council) (1978).'Concept of Uranium Resources and Producibility', National Academy of Sciences, Washington.

41. Holland, D. and U. Peterson (1981).'Element Dispersion, Element Concentration and Ore Deposits', in B. Skinner 'Economic Geology - Seventy-Fifth Anniversary Volume 1905-1980', Texas: Economic Geology Publishing Company.

INTEGRATED ASSESSMENT OF ENERGY-OPTIONS FOR CO_2 REDUCTION

T. Kram, P.A. Okken
Energy Study Centre
Netherlands Energy Research Foundation
P.O. Box 1
1755 ZG PETTEN, The Netherlands

ABSTRACT. Energy technology options for CO_2 reduction are evaluated in a process-oriented dynamic national costs minimizing LP-model of the Dutch energy system. To identify cost-effective CO_2 reduction strategies two scenarios are calculated with a 50% reduction of CO_2 emissions in 2020. CO_2 reduction sharpens the contrast between both scenarios: the nuclear supply-oriented all-electric strategy in the "Trend" scenario, versus the demand oriented gas strategy in the "Green" scenario. Some options are chosen in both scenarios to reduce CO_2. It is recommended to direct R,D&D-policy to these "robust" options, which include energy conservation, materials recycling, renewable energy and high efficient gas conversion technologies. Accounting for indirect CO_2 emissions and non-CO_2 greenhouse gases from the energy system is not expected to change the optimal reduction strategies.

This scenario study was sponsored by the Ministry of Economic Affairs (EZ) and by the Netherlands Agency for Energy and Environment R,D&D Management (NOVEM).

1. INTRODUCTION

Many options are available to reduce CO_2 emissions from the energy system. It is difficult to choose; e.g. options may be more expensive, not yet available or in conflict with each other. Analysis of options in a consistent framework of fuel prices, environmental constraints, demographic, economical, industrial and technological developments, is required to identify cost-effective CO_2 reduction strategies.

The Energy Study Centre operates various models to support national energy policy (Ministry of Economic Affairs) and energy research planning (NOVEM). These models are more sophisticated than the top down long term simulation models for greenhouse effect assesment, based upon aggregate econometric parameters (e.g., energy intensity of an economic system). On the other hand these models include more advanced technologies than the bottom up accounting models used by planning agencies for short term energy projections (8).

In this study a national economic optimizing dynamic LP model (MARKAL) was used. The model includes detailed data for the Dutch energy system; fuel prices; investment costs, availability, efficiency of several hundred different energy technologies; emission coefficients and emission abatement techniques. The model is used with exogenous national maximum allowable emissions ("bubble" concept). Changes in the energy system (fuel switch, recycling, emission abatement, changes in the energy technology mix, energy conservation) can compete to reduce the total national emission below the exogenously imposed maximum (the concept of "integrated" emission reduction). Ministers of the Environment from several countries (including Canada and the Netherlands) agreed in Toronto in 1988 on recommendations to reduce CO_2 emissions from the current level by 20% in 2005, and addressed the possible need for a further reduction of CO_2 emissions by 50% in a later stage. This paper reports on calculations to identify costoptimal CO_2 reduction strategies, in case the Toronto environmental policy recommendations were to become effective (1).

The model is dynamic and covers the period 1980-2020. This "perfect foresight" reflects the activities of policy-makers as they are assumed to "know the future". Dynamic modelling is of vital importance for testing structural changes in the energy system under environmental constraints. Some new energy techniques are supposed to become available at certain points in the future. This reflects the impact of ongoing energy and environmental R&D. These new speculative techniques include e.g.,: small scale absorption heatpumps, fuel cells, large off-shore wind turbines, underground coal gasification, electric or hydrogen vehicles (see table 3). The model enables testing of such new techniques in a future energy system under various constraints ("technology assessment"). Few institutional and market barriers are incorporated, to fully assess the potentials of (new) energy technologies. On the other hand upper bounds are imposed upon the rates of market penetration for new energy technologies, and there are lower bounds to ensure that older technologies will not be phased out too rapidly. In general researchers tend to introduce an optimistic view in their cost estimates and market assessments for new energy technologies. These optimistic estimates have been carefully evaluated (and adjusted if necessary) before use in this study (1).

General scenario characteristics

Scenario	TREND	GREEN
Industrial development	Export-oriented	Self sufficiency
Road Transport	Increasing	Decreasing
Environmental con-straints (NO_x, SO_2)	Mild (40% red.)	Severe (80% red.)
Nuclear energy	Unlimited	Phase out
Energy price increase	Moderate	Moderate
Economic growth rate	2,5 à 3%	2%
International context	Cosmopolism	Solidarity
	No CO_2 constraint in base case	No CO_2 constraint in base case

Scenarios for the period 1980-2020 were used in the model calculations, which reflect different world-wide socio-economic developments. These include a "Trend" and a "Green" scenario. The economic growth rate would be nearly 3% in the Trend scenario, and approximately 2% in the Green scenario. The real discount rate is 5%. Crude oil and gas prices are supposed to double in 2020 from the current Dfl 9/GJ. Coal prices would be about half of crude oil prices.

The Netherlands is a densely populated highly industrialized country. Domestic natural gas is the most important energy source. 98% of the residential and commercial sector is connected to the natural gas grid. The Dutch winter climate is moderate: 3000 heating degree days per year, with no significant cooling demand in summer. Electricity production is based mainly on coal and gas. Approximately half of the present oil consumption is for international bunkers and nonenergy use (petro-feedstocks); the other half is for inland transport. Oil use in the residential sector and for electricity production is negligible.

Primary energy supply in 2020 would be 2300 PetaJoule in the Green scenario, compared to 3400 PJ in the Trend scenario. In the Green scenario nuclear energy is banned and energy use decreases. By contrast in the Trend scenario no limitation to nuclear energy would persist and industrial production and transport energy use would continue to increase on the next wave of international growth and trade.

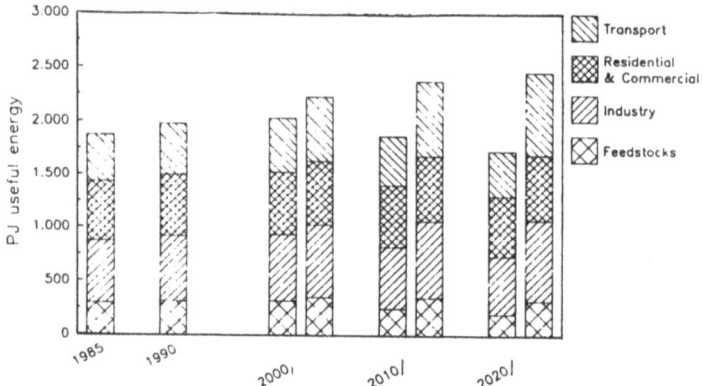

Figure 1. Energy demand by sector, base case scenarios
(Green: left; Trend: right).

2. CO$_2$ REDUCTION

Figure 2 shows in the upper lines the yearly CO$_2$ emissions
in the Trend and Green scenario base case. In accordance
with the recent national environmental policy master plan
(3) the CO$_2$ emission stabilizes in both scenarios by the
year 2000. As a sensitivity analysis the scenarios were
then recalculated with the recommended "Toronto" CO$_2$ constraint: 20% reduction in 2005, 50% reduction in 2020, compared to the current (1988) CO$_2$ emission. In the high
growth Trend scenario more CO$_2$ reduction would be needed,
compared to the Green scenario, to meet the Toronto recom-
mendation.

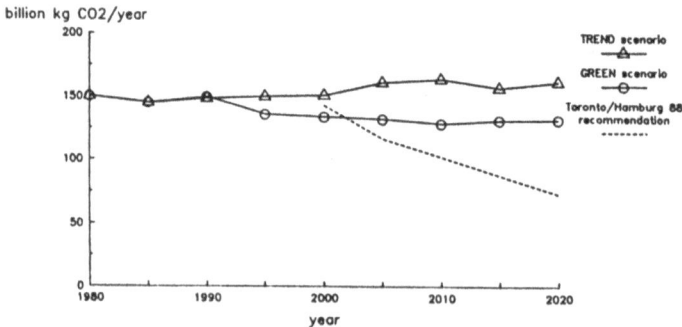

Figure 2. National CO$_2$ emission (excluding international
bunkers and petrochemical feedstocks) Trend and Green scenario; Toronto-recommendation for CO$_2$ emission reduction

The costs of CO_2 reduction are calculated by comparing both scenarios model-runs before and after the "Toronto" CO_2 constraints (table 1).

TABLE 1.   Costs of national economic optimal CO_2 emission reduction (1)

	Green scenario	Trend scenario
Cumulated CO_2 reduction (10^9 kg)	725	1364
Costs (10^9 Dfl)[a]	30,5[b]	48,8[c]
Dfl/ton CO_2	42	36

[a]   Dutch guilders 1988, undiscounted (1 Dfl o $ 0,5). Costs by deducting total system costs in scenario runs before and after CO_2 constraint (50% reduction in 2020).
[b]   Investment, operation, maintenance and salvage costs: 19,3; fuel costs: 11,2.
[c]   Investment, operation, maintenance and salvage costs: 101,4; fuel costs: -52,6.

The overall difference in costs per kg CO_2 reduced appears to be small. Yet a CO_2 constraint tends to sharpen the contrast between the nuclear supply-oriented all-electric strategy in a Trend scenario, and the demand-oriented gas strategy in a Green scenario. This fits well into the general scenario context. The identified costs of 50% CO_2 reduction (Dfl 30 .10y) would consume less than 0.5% of the Dutch GNP; this cost figure is of the same order of magnitude as the costs of 70% NOx/SO_2 reduction as committed in the current national environmental policy (3).
CO_2 reduction can be accomplished in various ways:
- fuel switching;
- increasing energy conversion efficiency;
- savings on energy end-use;
- recycling of carbonaceous materials.
These groups of measures are all available in the model used. The impact of the various CO_2 reduction measures are addressed below.

3. FUEL SWITCHING

Fuel switching is an important option to reduce CO_2 emissions. The specific CO_2 emission from coal burning is nearly twice as much compared to natural gas (table 2). Emissions from nuclear and renewable energy options are negligible.

TABLE 2. CO_2 emission coefficients (kg CO_2/GJ)[a] for various energy carriers (6)

Coal	94
Oil (gasoline, diesel)	73[b]
Gas	56
Uranium	0
Renewables	0

[a] GJ = Giga Joule primary energy, at lower heating value, combustion of fuels. Indirect emissions excluded.
[b] CO_2 emissions from refineries (appr. 5 kg CO_2/GJ oil products output) are also accounted in the model.

Fuel switching is a structural change in the energy system. Its potential can be properly assessed in a dynamic model. In the Trend scenario the model shifts from coal to nuclear power stations to meet the 50% reduction required in the CO_2 sensitivity analysis. Nuclear energy is not allowed in the Green scenario, under CO_2 constraints the model switches from coal to gas power stations. In both scenarios the contribution from renewable energy is increased (figure 3).

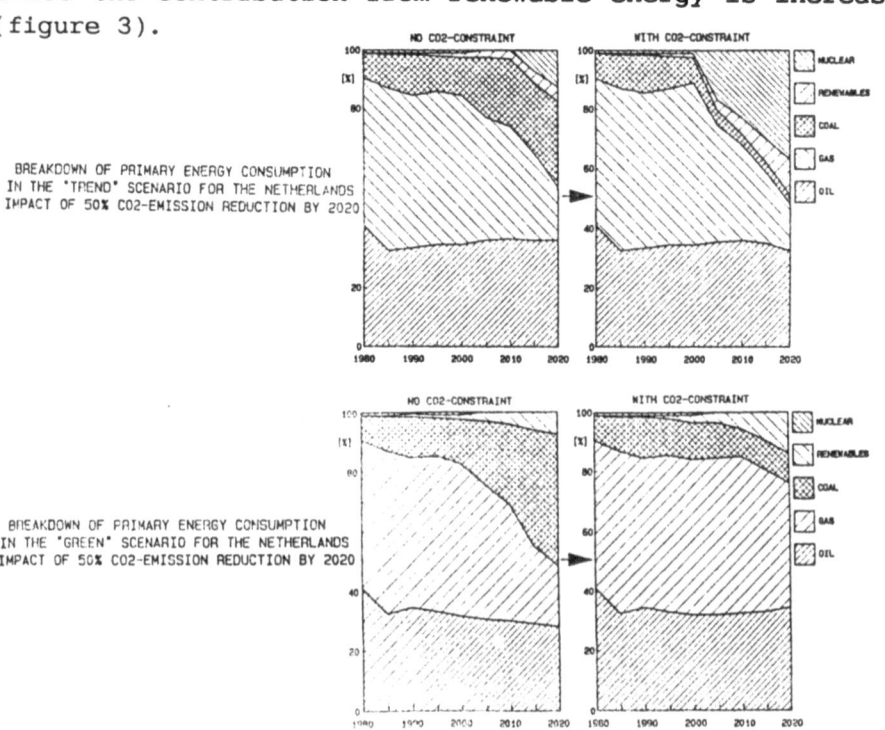

BREAKDOWN OF PRIMARY ENERGY CONSUMPTION IN THE "TREND" SCENARIO FOR THE NETHERLANDS IMPACT OF 50% CO2-EMISSION REDUCTION BY 2020

BREAKDOWN OF PRIMARY ENERGY CONSUMPTION IN THE "GREEN" SCENARIO FOR THE NETHERLANDS IMPACT OF 50% CO2-EMISSION REDUCTION BY 2020

Figure 3. Primary energy supply, before and after CO_2 constraint

Using nuclear energy instead of fossil fuels reduces CO_2 emissions. Under CO_2 constraints the nuclear capacity in 2020 in the "Trend" scenario is increased from 5 to 15 GWe. Investment costs in nuclear power plants, including the first core, are Dfl 4.100/kWe (1 Dfl o $ 0,5).

The zero specific CO_2 emission from nuclear generated electricity opens new markets for electricity, which would otherwise be served by gas or oil. Electric city cars (figure 8) and electric heatpumps in the residential and commercial sector are introduced under CO_2 constraints. Electric vehicles with overnight recharging improve the electricity load pattern for base-load nuclear power stations. The share of electricity in energy end-use shows a remarkable increase; the first steps towards an "all electric society" (figure 4). By contrast in the nuclear phase-out "Green" scenario the share of electricity decreases as a result of extra electricity conservation efforts under CO_2 constraints (figure 7).

Current Dutch nuclear capacity is 0,5 GWe, on line from 1968. Decisions to build new nuclear power plants have been delayed several times between 1970 and 1989. The question whether a nuclear revival to 5 or 15 GWe would be feasible in The Netherlands is not addressed he CO_2 emission reduction with no limitation to nuclear energy.

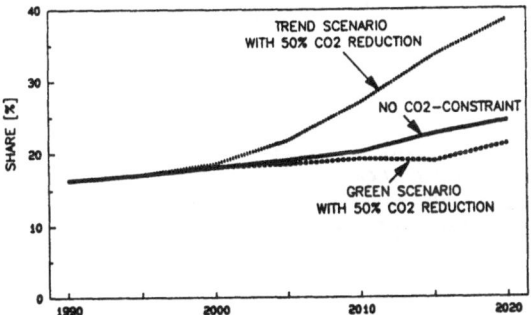

Figure 4. Electricity share in energy end-use

Natural gas is a domestic energy source. In total gas satisfies 52% of the Dutch primary energy requirements. Specific CO_2 emissions from gas are lower compared to oil or coal (table 2). Moreover gas can be used in high efficient gas conversion techniques. Under CO_2 constraints in the Green scenario coal power stations are replaced by gas combined heat & power (CHP) plants. In addition under CO_2-constraints gas is introduced in the transport sector. The declining share of natural gas in primary energy requirements in both scenarios from the present 52% to 20% in 2020 (figure 5), due to the depletion of domestic gas reserves and a switch from gas to coal as power station fuel, would not occur in the Green scenario under CO_2 constraints. Sig-

nificant quantities of natural gas would have to be impor-
ted, e.g., from Norway or the USSR. This fits well within a
possible general world context in which natural gas dominates energy supply, and could serve as a transient fuel to
delay greenhouse warming. Methane leakages from natural
gas use (in The Netherlands less than 1%) are not expected
to contribute significantly to global warming (7).

Substitution of renewable energy sources for fossil
fuels is a way to reduce CO_2 emissions (table 2). In both
scenarios the share of renewable energy in primary energy
supply is increased under CO_2 constraints (see figure 3).
The potential for renewable energy in The Netherlands is
relatively small, due to the high population density and
unfavourable climatic conditions.

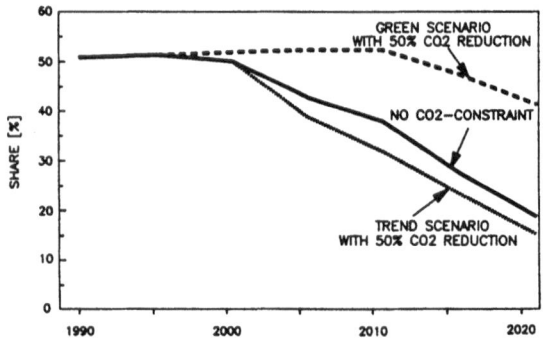

Figure 5.   Natural gas share in primary energy use

The main contribution is from wind, biogas and geothermal.
In addition solar PV cells are introduced in the Green scenario; solar dryers and bio-ethanol are introduced in the
Trend scenario. The contribution of hydro power and wood is
small (1).

4. ENERGY EFFICIENCY

Increasing the efficiency of energy conversion reduces CO_2 emissions. In the Green scenario a significant part
of the CO_2 reduction stems from high efficient gas conversion techniques: combined heat and power units, gas-en-
gines, gasturbines, condensing boilers, heatpumps and fuel
cells. Figure 6 illustrates the increase of fuel cells and
heat-pumps under CO_2 constraints. The mild winter climate,
the length of the heating season, the absence of a significant cooling demand and the nationwide availability of natural gas provide good opportunities for gas-fired heatpumps in the Netherlands.

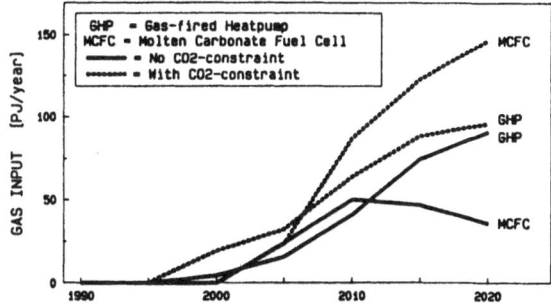

Figure 6. Fuel cell and gas heatpumps. Green scenario, impact of CO_2 constraint

In most studies on the greenhouse issue (2) energy end-use conservation is regarded as the most effective option to reduce CO_2 emissions. In our study economically attractive energy conservation options (e.g. thermal wall insulation, complete market penetration of 10 W light bulbs and high efficiency refrigerators) have already been included in the demand projections (3). The remaining potential and costs for energy conservation are difficult to assess. In the model calculations a tentative extra 30% energy end-use conservation in all sectors (except the transport sector) was assumed to be feasible for the opportunity costs of energy supply in 2020. This is a simplification, to detect the most attractive energy conservation sectors. Figure 7 illustrates the penetration rate of the "30% extra conservation" under CO_2 constraints in two aggregated sectors: space heating and electricity. Extra energy conservation in space heating is chosen in both scenarios. Extra electricity conservation is not chosen in the Trend scenario under CO_2 constraints due to the high share of nuclear energy, revealing a potential conflict between nuclear

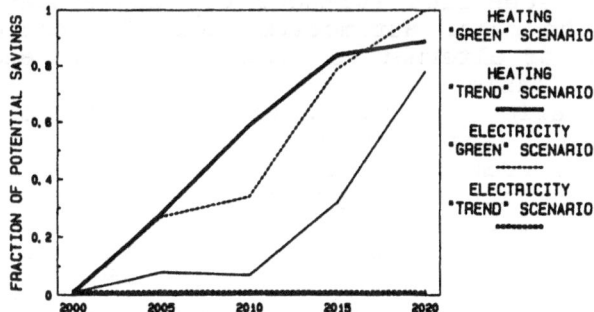

Figure 7. Penetration of extra energy end-use conservation in different sectors under CO_2 constraints.

electricity supply and conservation. In the Green scenario electricity conservation is a most attractive option to reduce CO_2 emissions.

5. RECYCLING

Recycling of carbonaceous materials reduces CO_2 emissions. Two options are considered: re-use of plastics and re-refining of waste lubricants. At present waste burning with energy recovery accounts for 35% of municipal solid waste treatment in The Netherlands. Environmental policy tends to increase the amount of waste being burned (3). The CO_2 emission coefficient of municipal solid waste (MSW) as an energy carrier is 100 to 150 kg/GJ, dependent on the plastic and rubber content (9 to 16% by weight). Re-use of plastics, instead of burning, reduces CO_2 emissions. Considerable amounts of waste lubricants are burned for horticulture and garage heating. Re-refining lubricants to new products could reduce CO_2 emissions. Other recycling options and materials substitution were not yet considered in this study although they could have a significant effect on CO_2 emissions.

6. TRANSPORT

The current share of road transport in The Netherlands CO_2 emission is 15%. In the Trend scenario transport energy use (and its associated CO_2 emissions) continues to increase. In the Green scenario energy use in road transport decreases to a pre-1980 level as a result of environmental policy to change the modal split: a shift from private to public transport (bus, train) and bicycles. The current modal split of road passenger kilometre production in The Netherlands (excluding walking) is: car 0.78; bus 0.06; train 0.07; bicycle 0.09.

In the base case scenarios (without CO_2 constraint) oil continues to be the major automotive fuel. To reduce CO_2 emissions a number of alternative automotive fuels and vehicle options are available for the model.

In the Trend scenario electric vehicles and biofuels (ethanol, vegetable oil) are introduced under CO_2 constraints. CO_2 emissions from bio-fuels are supposed to be zero in a sustainable agricultural practice. Also natural gas use is expanded, for CNG vehicles and for methanol production. The share of vehicles running on alternative non-oil fuels (biofuels, natural gas, electricity) would increase to 30% in 2020 in the Trend scenario under CO_2 constraints (figure 8). In the Green scenario however, no major fuel adjustments would be called upon under CO_2 constraints. The reduced amount of CO_2 and the share of the transport sector in the Green scenario are lower compared with the Trend scenario. This suggests that a policy to reduce CO_2 emissions by changing the "modal split" can avoid the costly introduction of alternative automotive fuels (or vice versa).

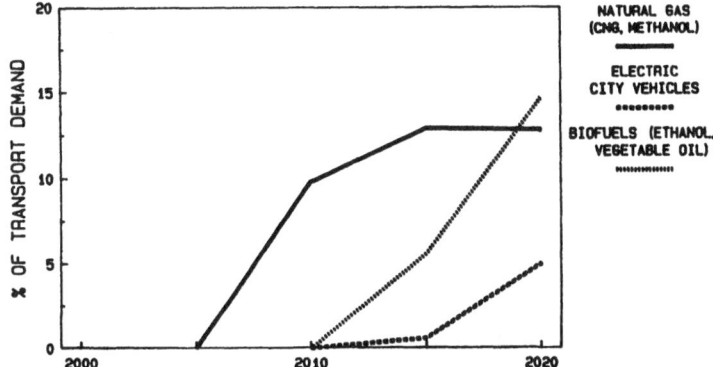

Figure 8.   Introduction of alternative fuels in the transport sector, Trend scenario under CO_2 constraints

7. R&D POLICY IMPLICATIONS

In general the Toronto CO_2 constraint sharpens the contrast between both scenarios. In the "international growth" Trend scenario nuclear energy, alternative automotive fuels and electric heatpumps are chosen by the model to reduce CO_2 emissions. Whereas in the "self sufficiency" Green scenario electricity conservation and combined heat and power become more attractive. Some options are attractive in both scenarios under CO_2 constraints. Table 3 presents a selection of energy technology options in the MARKAL-model. Options chosen by the model under CO_2 constraints both in the Trend and in the Green scenario are regarded to be "robust". It is recommended to direct energy policy and R&D efforts to these robust options. This can be looked upon as paying for "insurance policies" to reduce the risk of being caught unaware by an external CO_2 constraint. Major robust options are energy conservation, materials recycling, renewable energy (biogas, geothermal, wind) and high efficient gas conversion technology (CNG, fuel cell, heatpumps).

TABLE 3.

Energy options under 50% CO_2 constraint (1)	TREND	GREEN	
Absorption heatpump, medium, gas, comm.	0	+	
Absorption heatpump, small, gas, res.	+	+	R
Biogas industrial waste water	u	u	

Biogas landfill or manure	+	+	R
CNG bus/truck	+	+	R
CNG car	0	0	
Coal boiler/AFBC	-	-	
Coal gasification	-	-	
Coal power station	-	-	
Combined heat and power	-	+	
Condensing gas boiler	+	+	R
District heating	-	+	
Electric battery module car	0	0	
Electric city car	+	0	
Electric bus/trolley	+	+	R
Electric heatpump water heater	+	-	
Electric heatpump, space heating	+	0	
Electricity conservation	0	+	
Energy conservation, industry	+	+	R
Energy conservation, space heating	+	+	R
Ethanol/vegetable oil car/truck	+	0	
Fuel cell	+	+	R
Gas-engine combined heat and power	-	+	
Gas-engine heatpump	+	+	R
Gas STAG/CHP	-	+	
Gasturbine industrial CHP	-	+	
Geothermal	+	+	R
Hydrogen car/bus	0	0	
Hydrogen production by electrolysis	0	0	
Hydropower	u	u	
Hydropumped electricity storage	+	0	
Low energy building	+	+	R
Methanol car	0	0	
Methanol bus/truck	+	u	R
M-33 truck	0	0	
Methanol/synfuel production from coal	0	-	
Methanol/synfuel production from gas	+	+	R
Municipal solid waste burning	-	-	
Nuclear power station, LWR	+	0	
Oil automotive fuel, gasoline, diesel, LPG	-	0	
Solar drying, industry	+	0	
Solar electricity, PV-cell	0	+	
Solar water heater	0	0	
Stirling engine	0	0	
Underground coal gasification	-	-	
Windturbine land-based	u	u	
Windturbine off-shore	+	+	R

+ = Increased under 50% CO_2 constraint   0 = No change or not attractive - = Decrease u = Upper bound R = Robust option

The increase of worldwide CO_2 emissions is to be expected in developing countries and centrally planned econo-

mies. From this perspective R&D can also be directed towards options which are applicable in developing countries
(e.g., PV-cells) or in centrally planned economies (e.g.
new and improved gas technology).

8. INTERNATIONAL COMPARISON

The IEA/MARKAL-model has been applied in other countries to identify cost-optimal CO_2 reduction strategies. In
Japan CO_2 emissions in 2020 in a Trend scenario were reduced by 25% (which would be a 10% reduction compared with
the current level). Nuclear energy, electric heatpumps,
electric vehicles, PV-solar cells and geothermal are chosen
by the model in the Japanese energy system to reduce CO_2
(4). These results correspond well with our findings for
The Netherlands (table 3).

As another example in Sweden a 50% CO_2 reduction was
calculated for the regional energy system of Jönköping. Gas
combined heat and power, residential oil and wood heating,
hydropower and a limited increase in energy conservation
are chosen by the model in the Jönköping energy system to
reduce CO_2 (5). Differences with our results (table 3) stem
from regional circumstances, e.g. the absence of a residential gas grid, the severe building codes for thermal insulation, the potential for hydropower and energy forests in
Sweden.

9. INDIRECT CO_2 EMISSIONS, BUNKERS AND PETRO-FEEDSTOCKS

International bunkers and petro-feedstocks (mainly
plastic production) represent a significant part of the
Dutch primary energy consumption (20%, the global average
is ca. 8%). This reflects the concentration of refineries,
chemical industries, harbours and airports in The Netherlands For global evaluations CO_2 emissions from sea-
vessels, airplanes and feedstocks have to be accounted. In
this paper however, only CO_2 emissions from the Netherlands
territory subject to national energy policy are calculated,
at present 148 Mt CO_2/y (figure 2). CO_2 emissions from international bunkers and petro-feedstocks were excluded, although these would be equivalent to 23 and 13.10y kg
CO_2/year at present (1985). Inland burning of petro-feedstocks (MSW, lubricants) is included in the Dutch CO_2 emissions.

A related problem is posed by the CO_2 emission from
mining and transportation of fuels outside The Netherlands.
These "indirect" emission (table 4) arise at coal mines,
oil-well flare gas, overseas coal and oil transport, uranium mines and uranium enrichment.

The figures in table **4** represent the current Dutch situation. Indirect CO_2 emissions are one to two orders of magnitude lower than the direct CO_2 emission (table 2). Accounting for indirect emissions is not expected to change the cost-optimal CO_2 reduction strategies. The same applies to CO_2 emissions from production of limestone (used as flue gas desulphurization agent in coal power stations) and CO_2 emissions during building and decommissioning of power plants. By using energy analysis methods, these "building" CO_2 emissions were identified as negligible in most cases (6).

TABLE 4.   Indirect CO_2 emission coefficients (kg CO_2/GJ)[a] for various energy carriers in The Netherlands (6)

Coal	6
Oil	3
Gas	0,3
Uranium	3

[a]   GJ = GigaJoule primary energy, at lower heating value (fuels) or at thermal energy generated in a light water reactor (uranium).

10.   NON-CO_2 GREENHOUSE GASES

Other non-CO_2 greenhouse gases include methane (CH_4), carbon monoxide (CO), nitrogen oxides (NOx), nitrous oxide (N_2O), ozone (O_3) and chlorofluorocarbons (CFCs). These non-CO_2 greenhouse gases are to some extent interrelated with the energy system. E.g. CHd (natural gas leakages and CO-emissions) tropospheric Oc (from NO_x and traffic exhaust gases), N_2O (from catalytic car exhaust gas cleaning), CFCs (from refrigerators and heatpumps). Uncertainties about emission coefficients and atmospheric chemistry are presently too large to justify a detailed model calculation to detect cost-optimal greenhouse effect reduction strategies. Accounting for CH_4+CO emissions is not expected to change the costoptimal reduction strategies by fuel switching (7), although it might influence the choice of energy conversion technologies. Prevention of CH_4 leakage (from gas grids, coal mines and landfills) and prevention of CO emissions (from inefficient combustion processes and traffic exhaust gases) should be given high priority.

These typical "double-effect" options both increase energy efficiency and decrease CH_4 atmospheric concentrations.

11. FURTHER CO$_2$ REDUCTION

The calculations reported in this paper are based upon a 50% reduction of CO$_2$ emission. Some scientists and politicians say a larger CO$_2$ reduction will be necessary in rich countries, like The Netherlands.

To detect energy technologies which might become competitive under a more severe CO$_2$ constraint, benefit/cost-calculations have been performed. Figure 9 illustrates as an example the benefit/cost-ratio of low-CO$_2$ energy technologies in the passenger car system, "bubbling under" the market in the initial Trend scenario. The benefit/cost-ratio = 1 represents the "horizon" of penetrating techniques. Thus in the Trend scenario with 50% CO$_2$ reduction the electric city car (EC) penetrates the market in 2010. So does bio-ethanol (BE) produced from European agricultural surplus areas in 2015 (see also figure 8). The other low-CO$_2$ alternatives (EBM, Hb) improve their benefit/cost-ratios, but do not penetrate. Figure 9 suggests the electric battery module car (EBM) is likely to be a better alternative than the hydrogen car (Hb) if the need for further CO$_2$ reduction emerged, and EC/BE reach their upper bounds. However this calculation involved high purity hydrogen produced by electrolysis; other hydrogen production techniques (e.g. from natural gas or from a combined coal gasification/CO$_2$ removal process) might reduce production costs and further improve the benefit/cost-ratio of hydrogen cars.

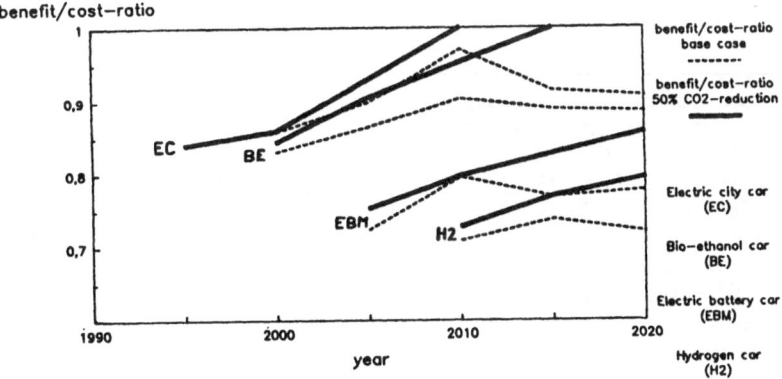

FIGURE 9. Benefit/cost-ratio for selected passenger car non-oil alternatives. Trend scenario, impact of CO$_2$ constraints

12. CARBON TAX

A carbon tax on fossil fuels to combat the greenhouse effect has often been debated in energy and environmental policy (2,3). One of the questions might be: which carbon tax is required to gradually reduce CO_2 emissions by 50% in 2020.

A minimum required carbon tax can be deduced from the model calculations reported in this paper. In our view the expenditure to reduce CO_2, in a period in a scenario, must be earned via a "carbon tax" on the remaining fossil fuel use in the same period in the same scenario. The CO_2 reduction and the remaining CO_2 emission from fossil fuel use in both scenarios are presented in figure 2 as the area between the solid and dashed line resp., the area below the dashed line. The minimum required carbon tax is calculated from the cost of national optimal CO_2 emission reduction (table 1) and the remaining CO_2 emission (figure 2).

Such a carbon tax is only a small (5 to 10%) increase in the projected crude oil and gas price, and a moderate (20 to 30%) increase in the projected coal prices (1). The calculated carbon tax falls within the range of differences between various "official" energy price projections. On a end-use base (including indirect CO_2 emissions, table 4) the carbon tax would be Dfl 0.027 to 0.042 per ms^3 natural gas, Dfl 0.045 to 0.071 per liter gasoline, Dfl 0.041 to 0.063 per kg coal, and Dfl 0.012 to 0.019 per kWh electricity supplied from a modern IGCC coal fired power plant.

TABLE 5. Minimum required carbon tax for 50% CO_2 reduction

Scenario	GREEN	TREND
CO_2 reduction costs (10^9 Dfl)[a]	30.5	48.8
Remaining CO_2 emission (10^9 kg)[a]	2019	2078
Carbon tax (Dfl/ton CO_2)	15.1	23.5
Carbon tax for gas (Dfl/GJ)[c]	0.8	1.3
Carbon tax for oil (Dfl/GJ)[c]	1.1	1.7

[a] See table 1, guilders 1988, undiscounted, 1 Dfl o $ 0,5
[b] Period 2000-2020 cumulated, see figure 2
[c] See table 2

This carbon tax to reach 50% CO_2 reduction in 2020 (table 5) seems to be a minimum requirement for several reasons:
- The model implicity assumes a "perfect" market economy with perfect foresight. In many markets energy consumption is known to be highly in-elastic and short-sighted. Subsidies, energy efficiency standards and government investments will definitly be needed to ensure the national optimal CO_2 reduction strategy (either in the

Green or in the Trend scenario) as identified in this
paper;
- The raised carbon tax on fossil fuels, and the re-al-
location of funds within the energy sector, imply an
effective governmental "clearing house" mechanism. However in most countries fuel taxes are mainly used to
decrease national budget deficits;
- The calculated carbon tax is a mean for the remaining
fossil fuel consumption in the period 2000-2020. During
this period the carbon tax should progressively be increased to maintain a low CO_2 emission level after 2020.
The lower level of fossil fuel consumption (e.g., in the
Trend scenario, see figure 3) might result in lower
energy prices, thus increasing energy demand and CO_2
emissions;
- CO_2 reduction results in part from new energy technologies, now under development. R&D funding is not expli-
city included in the projected costs of these new energy
technologies. There is also a risk that some of these
technologies will not become available in time, despite
extensive R&D funding.
For these reasons we feel a somewhat higher carbon tax in
the period 2000-2020 is needed, if Dutch CO_2 emissions are
to be reduced by 50%.

13. REFERENCES

1. Kram, T. and P.A. Okken (1989). Two low COb energy scenarios for the Netherlands, Proceedings IEA/OECD expert
   seminar on energy technologies for reducing greenhouse
   gas emissions, Paris, France.
2. Nota Energiebeleid (1974). Ministry EZ, The Hague, The
   Netherlands.
3. Nationaal Milieubeleids Plan (1989). Ministries VROM,
   EZ, L&V, V&W, The Hague, The Netherlands.
4. Koyama, S. (ETL) and S. Yasukawa (JAERI) et.al. (1988).
   A preliminary energy-environment analysis: Part 1 and 2,
   Proc. of a joint ETSAP/IIASA-workshop BNL, Upton, New
   York, USA.
5. Larsson, T. and C-O. Wene (1988). Robust energy systems:
   a Swedish community as an example, Chalmers University,
   Gerg, Sweden.
6. Blok, K., J. Bijlsma, S. Fockens and P.A. Okken (1988).
   CO_2 emission coefficients for fuels and energy carriers
   in The Netherlands (in Dutch), Utrecht University/ESC,
   Petten, The Netherlands, ESC-WR-88-12.
7. Okken, P.A. and T. Kram (1989). CH_4/CO-emissions from
   fossil fuels global warming potential. ESC, Petten, The
   Netherlands, ESC-WR-89-13.

8. Okken, P.A., J. Rotmans, R.J. Swart (1989). The green-house effect: uncertainty dilemma between science and policy, and its effect on scenario development. Presented at IEA/IPCC expert group meeting on methodologies and analytical tools. RIVM, Bilthoven/ESC, Petten, The Netherlands, ESC-WR-89-07.

CARBON DIOXIDE AND POLICY OPTIONS

W. VAN GOOL
Energy Science Project
State University Utrecht
Croesestraat 77A
3522 AD Utrecht, The Netherlands

ABSTRACT. The greenhouse effect and its relation to the carbon dioxide emission has been studied intensively in the last two decades. It is beyond doubt that a continued economic growth will lead to an increase of the average world temperature, if the energy supply pattern remains unchanged. The climate will change, but consequences on a regional scale cannot be specified. Economic models are not good enough to predict the moment that the carbon dioxide concentration will have doubled as compared to the pre-industrial value. Chemical methods for using carbon dioxide as feedstock in the production of durable consumer goods are inadequate due to the scale of the carbon dioxide emission. International co-operation is necessary to reduce the emissions, but the fact that the climate change could benefit some countries makes it difficult to reach an international agreement. It is possible to describe in detail the energy structure of a society with low carbon dioxide emission, while maintaining today's activities. Many decades are necessary to implement the changes. The short term structure of many political systems and the low social acceptance of certain features of the low-emission society suggest that society had better prepare itself for certain consequences of the change of climate.

1. INTRODUCTION

Carbon dioxide emission and its influence on climate have been studied intensively during the last two decades. Many reviews dealing with several aspects of the problem are available [1-7]. In 1982 a Carbon Dioxide Information Center was established in Oak Ridge (Tennessee, USA). In the first national energy research plan in The Netherlands (1975) the carbon dioxide problem was pointed out as the limiting factor for a continued growth of the use of fossil fuels.

In spite of all this information it has taken 15 years before the problem appeared on the political agenda. This is one example of the long lead-time between scientific recognition and political interest. One of the reasons for this delay is that the carbon dioxide issue is not much appreciated by those who oppose the use of nuclear energy in the electricity production.

It is noted only about one half of the greenhouse effect is due to carbon dioxide. Since the gases causing the other half, such as CFKs (chlorofluorocarbons), are emitted to the atmosphere in much smaller quantities than carbon dioxide, the prospects to solve this part of the problem look much better than in the carbon dioxide case. Indeed, the amounts of carbon dioxide emitted, are staggering. (See Diepstraten's paper, this conference).

In this paper the effects on climate are summarized. The problem of the time to double the concentration of dioxide in the atmosphere relative to the pre-industrial value is discussed in relation to economic models. It is shown that backcasting (normative modelling) rather than forecasting offers better possibilities to develop a low carbon dioxide policy. Structural aspects opposing the implementation of such a policy are mentioned.

The ideas developed in this paper are based upon the fact that the carbon dioxide problem coincides with the energy problem to a large extent. The limited results of all work carried out in the last decade in the world to "solve" the energy problem suggest that the carbon dioxide problem cannot be "solved" within the political and social structures of our societies.

2. CLIMATE AND INTERNATIONAL CO-OPERATION

Scientist agree that a doubling of the carbon dioxide concentration in the atmosphere will lead to an increase of the average world temperature of about 4 degree Kelvin. (Doubling is relative to the pre-industrial value (280 ppm), leading to a concentration of 560 ppm. The present value is 350 ppm.). Now we are probably going through the first degree out of four. Four degrees seems moderate, but it would make Earth the warmest in the last 150,000 years.

The consequences for the climate cannot yet be determined in detail: it is especially impossible to predict, for example, the pattern of hurricanes and other local extreme phenomena.

In connection with the desirability to develop international co-operation it is relevant that climate changes might make agriculture possible in certain parts of the world where it is now impossible. This will not stimulate countries in those areas to participate in drastic international efforts to decrease the carbon dioxide emissions.

3. MODELLING THE DOUBLING

Many studies have been devoted to the problem when the doubling of the carbon dioxide concentration will be reached.

The rate of increase of the concentration depends upon many factors, such as e.g., the change of the size of the world population, the economic growth in the developing countries, and the relation between energy use and economic growth [4, 8 - 11]. The value of this long term economic modelling is very limited [13]. Changing the parameters leads to a wide variation in doubling time. A moratorium for long term economic modelling has been suggested [12].

Backcasting -instead of forecasting- is more useful for developing a policy. In backcasting one sets a technically feasible objective in a certain year or period in the future and one establishes what must be done to arrive at that objective from the present situation. It is not suggested that backcasting is an easy way: selecting a year and selecting the technically feasible objectives require knowledge, wisdom and a great amount of help from several disciplines. Lovins [4] and Goldemberg [3] follow -more and less- this approach. Diepstraten (this conference) uses levelling of the carbon dioxide emission in The Netherlands in 2015 as the objective.

4. A POLICY OR NO POLICY?

The question whether or not to have a greenhouse policy is studied internationally [2]. Several lines of thought are being followed.
One approach is that the uncertainty about the consequences of the greenhouse effect are so large that immediate action is necessary. This idea fits into the nuclear energy discussion: one shall not use nuclear energy as long as there is any uncertainty about the long term consequences of this application. Obviously, using the same reasoning in these situations increases the contradiction: not using nuclear energy increases the uncertainty about the carbon dioxide effects.

A second reasoning is that curtailing the carbon dioxide emission faces so many difficulties that preparation for an expected climate change should be included in the policy.

The third line is obvious: nothing is sure about the future, so we should do nothing special now.

The final result is that one probably can compromise on the tie-in strategy: take measures favourable for the carbon dioxide issue only when they fulfil other desirable objectives. These measures should survive even when the carbon dioxide problem should "fade" or disappear from the political agenda. The tie-in strategy reflects strongly the recent experience with energy policy, where many policies, designed for eternity, did not survive one decade.

5. BLUEPRINT, BACKCASTING AND TIE-IN

Using the trends and ideas sketched in the previous sections it is possible to develop the guidelines for policy with respect to the carbon dioxide problem.

5.1 The blueprint

The advantage of the intensive energy research in the past decade is that the technical description of the low carbon dioxide future is not very difficult.

The largest contributions to the carbon dioxide emission come from the Household & Other Sector and the Industry Sector (Diepstraten, this conference). The sectors Transport Sector and the Electricity Production Sector contribute much less. So, any plan that does not attack the energy requirements of homes, buildings, and industry cannot solve the problem.

Furthermore, the mere size of the carbon dioxide emission is such that no single measure can solve the problem. Thus, both the Transport Sector and the Electricity Production Sector also have to contribute.

The contribution of the Electricity Sector is relevant for another reason: solving the problems in the Household & Other Sector and some of the problems in the Transport Sector require a larger application of electricity and therefore electricity production based upon non-fossil sources is essential to the solution.

I shall mention a number of the options:
- increase the insulation standard of homes and buildings and improve the ventilation system to such a level that electric heating becomes efficient (heat pumps)
- rearrange industrial productions in such a way that heat handling and heat integration decrease the energy requirement with 20 to 30% (heat exchange, cogeneration, heat pumps, heat engines) [14, 15]
- develop electric vehicles and other electric transport for local use in cities

- replace the fuels used for electricity production in the sequence coal --> oil --> natural gas --> non-fossil (nuclear, fusion, solar, wind, hydrothermal, tides, bio-mass)
- when fossil fuels still have to be used in large heating units, it is possible to use the carbon dioxide from the stack gas to produce syngas. Syngas can be used for many purposes, for example to make methanol for the transport sector and for the production of polymers.

Many of these technical options are or will be available. Many more options are being developed (ceramic materials for high temperature applications, fuel cells, and superconducting materials.), but it will increase their acceptability when plans are based upon facts rather than upon expectations. Considering the long term character of the planning, there is ample time to adjust to unexpected developments.

5.2 Backcasting and tie-in

The backcasting must be developed by taking into account the proper lead-times. The major task will be the development of an economic path. The economic aspects can be favourable for many parts of the plan when the normal depreciation of the investments can be used for the development.

Note that many new structures, buildings, and equipment will be required. A large fraction will have to be made in the "old" structure. Benefits with respect to the carbon dioxide emission are not obtained in the short term, but once the restructuring comes to an end, the benefits will be much larger than those which can be obtained with short term actions.

The development has to start with the tie-in strategy. Parts of the restructuring fulfil other objectives of the present policy plans. For example, the use of electric vehicles for city transport can reduce air pollution in the cities and improving the energy efficiency of homes and buildings will be beneficial in several ways.

Such a long term energy (epsilon) plan is technically and economically feasible.

5.3 The bottlenecks

Strongly incorporated in many societies are some bottlenecks that will prevent the execution of the epsilon plan.
Execution of the plan requires an effort that will continue for many decades. We have no organization that can enforce the execution, except the existing political structure. This structure, however, is based upon a short time struc-

ture, generally four to six years. Every new government has the right to change priorities. The long term continuity is not guaranteed.

One could think about a structure or organization above or besides the political institutions with sufficient power to guarantee the continuation of the execution of the epsilon plan. However, no self-respecting democracy accepts an organization without any political control, while that organization makes changes influencing the life of everybody.

Other bottlenecks originate from the fact that the increased energy productivity is obtained by an increasing use of electricity based upon non-fossil resources. Many groups reject that approach.

Probably the most devastating aspect of the epsilon plan will be the result when the plan is ready: nothing will show for it. Surely, there will be less carbon dioxide in the atmosphere than without the plan, but it is impossible to see somewhat less of something that is invisible. No official person can put a flag on the result of the plan. No politician can show the result to the voters.

The conclusion is that the plan will not be executed. It is better to prepare society for the coming climate change. This means, for example, increasing the height of dikes, and making cooling equipment. These things one can see and feel, just as the consequences of a hurricane.

6. REFERENCES

1 Clark, W.C. ed., (1982). Carbon Dioxide Review 1982, Oxford University Press.
2 Hilleman, B. (1989). 'Global Warming' in Chem. Eng. News 67(11) pp. 25-44.
3 Goldemberg, J. et al., (1985). 'An end-use oriented global energy strategy', in Ann. Rev. Energy 10, pp. 613-8.
4 Lovins, A.B. et al., (1981). Least Cost Energy: Solving the CO_2 Problem, Andover, Mass: Brick House.
5 Rotty, R.M., C. Masters, (1984). 'Past and Future Releases of CO_2 from Fossil Fuel Combustion', Inst. Energy Analysis, Oak Ridge, Tenn.
6 Edmonds, J., J.M. Reilly, (1985) Global Energy Assessing the Future, Oxford: Oxford University Press.
7 Edmonds, J., J.M. Reilly, (1985). 'Time and uncertainty: analytic paradigms and polic requirements' in W. van Gool and J.J.C. Bruggink (eds.,), Energy and Time in the Economic and Physical Sciences, North-Holland, Amsterdam pp. 287-313.
8 IIASA 1981. Energy in a Finite World: A Global Systems Analysis, Ballinger Cambridge, Mass.

9 WEC 1983. Energy 2000-2020: World Prospects and Regional Stresses, Graham and Trotman, London.
10 Ausubel, J., W.D. Nordhaus, (1983). 'A review of estimates of future carbon dioxide emissions' in Changing Climate, Natl. Acad. Sci., Washington D.C.
11 Reister, D.B. (1984). 'An Assessment of the Contribution of Gas to the Global Emissions of Carbon Dioxide', Final Report GRI-84/003, Gas Res. Inst., Chicago.
12 Keepin, B. (1986). 'Global energy/CO_2 projections', Ann. Rev. Energy 11, pp. 357-92.
13 Landsberg, H.H., Commentary to: Perry, A.M., 'Carbon dioxide production', in ref. 1.
14 Gool, W. van, R. Kümmel, (1986). 'Limits for cost and energy optimization in macro-systems', in Miyata Matsui (ed.,), Energy Decisions for the Future, Vol. I, pp. 90-106.
15 Groscurth, H.M., R. Kümmel, W. van Gool, (1989). 'Thermodynamic Limits to Energy Optimization', in Energy 14, pp. 241-58.